Solid-State
Power Electronics

by
Irving M. Gottlieb

Howard W. Sams & Co., Inc.
4300 WEST 62ND ST. INDIANAPOLIS, INDIANA 46268 USA

International Standard Book Number: 0-672-21574-8
Library of Congress Catalog Card Number: 79-62992

Printed in the United States of America.

Preface

Technologists are not immune to nostalgia which often titilates the minds of others. Those of us who have worked long in the field of electronics can recall the sentiments in fashion during the early era of the transistor. A then-prevalent attitude asserted that a clever achievement in the domain of applied science had indeed been made, but one obviously destined for limited applications. According to this outlook, one might expect to encounter transistor implementations only in toys and novelty circuits, and that its evolution was being nurtured only through articles in journals catering to experimental projects and in secluded laboratory investigations. However, those with technical insight made the concession that a new electronic device had surely been added to the engineer's arsenal—one that deserved recognition as an *adjunct* to the electron tube.

Enjoying the luxury of retrospect, we find that this type of thinking evolved during the few years following the introduction of the transistor. The era of crystal-set audio amplifiers, small relay drivers, and flea-power oscillators was short lived. As if germanium and silicon were, among other things, growth hormones, it was not long before solid-state projects escaped from power restraining levels of tens of milliwatts, from current limitations of a few milliamperes, as well as from frequency responses in the low audio range. Because of junction power transistors and a related device, the SCR, solid-state applications began to require heat sinks to enable the handling of tens of watts, and more, of power dissipation. Notwithstanding, one still heard "authoritative" voices declaring the solid-state devices to be fads, to be inherently unreliable, and to lack characteristics needed for serious engineering consideration.

Despite universal acceptance of solid-state devices, few appear to appreciate the extent to which solid-state capability has evolved. We casually accept the role of solid-state devices in several-hundred watt stereo amplifiers, but may manifest surprise to learn of their employment as the main control elements in traction vehicles. And, it may require a good measure of mental nimbleness to reconcile the gamut represented by electrical parameters in a transistor radio and those of the solid-state inverters that process electrical power at kilovolt and kiloampere levels. Even at lower power levels, concepts of power limitations of such devices as ICs and FETs often prove grossly erroneous, if not archaic.

Obviously, a treatise is *not* needed to reinforce any less-than-firm convictions that solid-state power circuits and systems are here to stay. It is, however, the author's strong conviction that a book depicting the wide range of applications for solid-state power electronics can render a useful service to the technical community. Accordingly, *Solid-State Power Electronics* has been written to provide a broad overview of devices, techniques, and circuits based upon solid-state technology. This should enable the experimenter, technician, engineer, and others to make implementations suited to specific needs. In addition, it is felt that those whose primary concern involves operation, maintenance, and servicing will also derive valuable insights from the extensive spectrum of material presented in this book.

The root source of information processed in a book of this kind is derived from the competently staffed and well-equipped research and development laboratories of the semiconductor firms. The assistance rendered by the following individuals and their respective firms is gratefully acknowledged by the author. W. C. Caldwell, Distributor Sales Administrator, Delco Electronics; Neil Cleer, Manager, Marketing Services, International Rectifier Corp.; Walter B. Dennen, Manager, News & Information, RCA Solid-State Division; Robert Dobkin, Director, Advanced Circuit Development, National Semiconductor Corp.; Forest B. Golden, P. E., Customer Engineering, General Electric Semiconductor Products Department; Dave Hoffman, Applications Engineer, Siliconix, Inc.; and Lothar Stern, Manager, Technical Information Center, Motorola Semiconductor Products, Inc.

IRVING M. GOTTLIEB

Contents

sistor Breakdown Characteristics — Thermal Cycling of Power Transistors — Selection of Transistor Types — Basic Types of SCRs — Applying SCRs and Triacs — Commonly Used SCR Triggering Circuits — Triggering Triacs — Things to Keep in Mind About Thyristors — A New Era in Solid-State Power

CHAPTER 4

CHAPTER 5

1

Basic Principles of
Solid-State Control of Power

WHAT IS POWER ELECTRONICS?

From its inception, the wonder of electronics was derived largely from the detection and control of very feeble power levels. Admittedly, this association stems from early days of wireless communication and from the then-awesome ability of electron tubes to deliver useful outputs from minute antenna currents. Even though giant transmitting tubes and industrial tubes of multikilowatt capability ultimately evolved, a general understanding prevailed concerning the differentiation one was obliged to make when dealing with radio reception and when dealing with so-called power engineering. It was more or less directly from *radio* that the more comprehensive term *electronics* came into being. Intentionally, or otherwise, electronics continued to remain suggestive of the processing of tiny signal levels.

With the further progress in semiconductor technology, devices and applications evolved which readily delivered power greatly exceeding that needed merely to actuate a meter, display, or small transducer. Moreover, these new devices could, by themselves, or in conjunction with solid-state circuitry, produce hefty outputs when actuated by low-power signals. What had emerged was a new state of the art not accurately described by the phrase, *power engineering*. Rather, aspects of both classifications clearly applied. Thus, we have the term *power electronics*.

It is only natural to ponder the boundaries of power electronics. In other words, at what power levels does it commence, and perhaps end? A partial answer resides in the allusion to *heat removal*. In a general way, we may assume we are dealing with power electronics when heat removal merits consideration. In practice, it is found that this situation prevails whenever a solid-state device must dissipate one watt or more. At ten watts, there is little question of such need. So, somewhat arbitrarily we may say that the beginnings of power electronics are in clear evidence at the three-watt power dissipation level. With regard to an upper limit, this is best left undefined since history suggests the hedged statement that "the sky is the limit."

WHERE IS SOLID-STATE POWER ELECTRONICS USED?

It sometimes happens that a meaningful answer to a question is provided by *another* question. It is appropriate to the subject of solid-state power electronics that we might ask where it is *not* applicable. Certainly, such a query would invoke but a small fraction of the applications listed in Table 1-1. The implication is that solid-state power electronics has an almost universal involvement in modern technology. Milestones on the path of electronics progress were marked by such advents as microwaves, television, semiconductors, and computers. And it appears abundantly clear that the present era largely involves the domain of solid-state power electronics.

WHAT IS UNIQUE ABOUT
SOLID-STATE POWER ELECTRONICS?

Much of the answer to such a query is surely well known to the reader. All of the desirable characteristics of semiconductor devices apply. That is, the small size, long life, low cost, easy application, low maintenance, and physical ruggedness of semiconductor devices are as attractive for power electronics as for other applications. But, the truly compelling feature of solid-state power devices is the comparative ease with which heat removal can be implemented. This is due to several factors. First, solid-state devices tend to have higher electrical efficiency than other electronic control devices so that less heat is generated within the device itself. (Tubes, in contrast, have high plate-cathode voltage drops. And, their "heaters" are well named, for much more energy is converted to heat than goes into thermionic emission of electrons.) Secondly, it is generally more practical to devise effective thermal hardware for a compact device fabricated of solid materials in intimate contact than it is for a large and, perhaps, irregularly shaped device in which the heat-generating elements are not thermally accessible. Thirdly, the operating temperature of silicon devices can be quite high—the early notion of a semiconductor device as being vulnerable to a few tens of degrees Celsius above room temperature is no longer valid.

Table 1-1. Some Application Areas of Solid-State Power Electronics

ADVERTISING	MOTOR CONTROL
AIR CONDITIONING	MOVIE PROJECTORS
AIRCRAFT	MOVING LIGHTS
ALARMS	NEON SIGNS
AMPLIFIERS	NIXIE DRIVERS
APPLIANCES	OVEN CONTROLS
AUTOMOBILES	PROJECTORS
BATTERY CHARGERS	PHONOGRAPHS
BLENDERS	PHOTOCOPIERS
BLOWERS	PHOTOGRAPHY
BOILERS	POWER LINE CONDITIONERS
BUOY LIGHTS	POWER SUPPLIES
BURGLAR ALARMS	RADAR
CHEMICAL PROCESSING	RADIO TRANSMITTERS
CLOTHES DRYERS	RANGES
COMMUNICATIONS	REFRIGERATORS
CONVERTERS	REGULATORS
CRANES	RELAY REPLACEMENT
DIMMERS	SANDERS
DISPLAYS	SEWING MACHINES
DOOR OPENERS	SERVO SYSTEMS
ELECTRIC BLANKETS	SOLENOID DRIVERS
ELECTRIC VEHICLES	SPACE VEHICLES
ELEVATORS	SWEEP CIRCUITS
ELECTROLUMINESCENT PANELS	SWITCHING
ELECTROMAGNETS	TELESCOPE CONTROL
ELECTRONIC IGNITION	TELEVISION CIRCUITS
ELECTROSTATIC PRECIPITATORS	TEMPERATURE CONTROL
FANS	THYRATRON REPLACEMENT
FLASHERS	TIMERS
FLOOR POLISHERS	TOOLS
FOOD MIXERS	TRAFFIC LIGHT CONTROL
FREQUENCY CONVERTERS	TRAINS
GAMES	ULTRASONIC PRODUCTS
GARAGE DOOR OPENERS	UNINTERRUPTIBLE POWER SUPPLIES
GAS APPLIANCES	UTILITY SYSTEMS
GRINDERS	VACUUM CLEANERS
HAND TOOLS	VEHICLES
HEAT CONTROL	VENDING MACHINES
IGNITION	VENTILATING EQUIPMENT
INDUCTION HEATING	WASHING MACHINES
INDUSTRIAL PROCESS CONTROL	WELDERS
INVERTERS	XEROGRAPHY
LATHES	ZERO-VOLTAGE SWITCHING
LIGHT CONTROL	
LATCHING RELAYS	
MACHINE TOOLS	
MAGNETS	
MAGNETIC RECORDING	

In addition to the thermal and other features of semiconductor devices, we find that the performance of solid-state power devices is admirable in other respects, too. Thus, one is not hard pressed to discover such devices with excellent linearity, or with speedy response. And, although the law of tradeoffs cannot be altogether evaded, its assertion is often quite benign for many practical implementations.

(A) Amplification with a vacuum tube.

(B) Phase control of load power with a thyratron.

(C) On-off load power control with a electromechanical relay.

(D) Saturable reactor used to control ac load power.

(E) Speed control of dc motor (Ward-Leonard system).

Fig. 1-1. Some nonsolid-state control methods.

For example, a transistor which can switch 500 volts at a 20-kHz rate and provide several amperes from a regulated power supply with a switching efficiency of over 90 percent is a mundane rather than an exotic item.

It would be presumptuous to imply that all older control devices (Fig. 1-1) have been rendered obsolete. It is true, however, that solid-state manipulation of power exhibits superiority in such a wide spectrum of applications that it merits study as a unique control technique.

BECOMING PROFICIENT WITH SOLID-STATE POWER ELECTRONICS

To be well-versed in the general concepts and practices underlying "ordinary" electronics is an excellent foundation for gaining proficiency with solid-state power electronics. However, power electronics differs from the more ordinary phase of the electronics art in that it requires stricter compliance with the basic postulates of electrical engineering. Such a statement may appear trite in the sense that *all* electronic phenomena are governed by the fundamental laws laid down by the physical sciences. In the world of practical experience, however, one often gets by quite well in dealing with ordinary electronic circuitry by paying scant heed to many things which assume importance when power levels are high. In power electronics, one is much less likely to enjoy the luxury of another "cut and try" remedy for unsatisfactory results. Whereas the penalty for violation of some basic electrical law may have previously been a glowing plate in a vacuum tube, semiconductor devices are relatively unforgiving. Not only are they vulnerable to destruction when their ratings are exceeded, but they often become involved in a "chain reaction" in which a number of other devices and components also suffer destruction.

Some of the matters which assume importance in power electronics are power factor, transients, the measurement and consequences of various waveforms, polyphase circuits, duty cycle, energy storage, switching phenomena, and the behavior of low-Q resonance. Although these topics also pertain. to ordinary electronics, they merit a more focused emphasis when working in the domain of power electronics. They will be discussed accordingly with this relevancy in mind.

Interpretation of Measurement and Specifications

Many misunderstandings and malperforming circuits occur in power electronics because of the misinterpretation of measurement data. For example, a large switching-type power supply will be ordered for a digital system. It is found that although the supply provides its rated voltage and current and provide excellent regulation, it causes false triggering of sensitive logic circuits. At first, even though the power supply is suspect, the trouble is elusive because ripple and noise from the supply are well within the manufacturer's specifications. Eventually it is discovered that these ratings are specified in terms of their rms value—a perfectly acceptable method, per se. However, some of the ripple and noise components are far from sinusoidal waveshape. So, instead of the ratio of peak to rms value being a mere 1.41, as with sine waves, this ratio may be five to one, ten to one, or even greater, as tends to be the case with switching spikes. And, unfortunately, logic circuits tend to be "trigger happy"—they respond to *peak,* rather than rms levels. Inasmuch as such transients are not always easily eliminated with filtering and bypassing techniques, it is always desirable to anticipate such possibilities in advance.

Another consequence of such short-duration transients is the actual destruction of semiconductor devices. One might contend that the energy content of very narrow spikes is negligibly low (as proven by the rms value). However, they may easily prove to be the "straw that breaks the camel's back." This does not imply that the destroyed devices were necessarily operating without adequate safety margin. Rather, it is a manifestation of the fact that these devices tend to be voltage sensitive in the sense that a momentary penetration of their soa (safe operating area) boundaries suffices to trigger catastrophic destruction. And such destruction is often possible even after the designer has taken suitable precautions to protect against the "worst possible" overvoltage that could occur under ordinary operating conditions. The proper protection against destructive transients is, of course, an awareness of the various measurement parameters of voltage and current levels, particularly as they apply to nonsinusoidal waves.

In the example just cited, the ideal situation would prevail if the vendor would provide specifications in *both* rms and peak values. Additionally, the user should be informed of the *type* of instrumentation employed to provide the measurements. Among other things, instruments, such as oscilloscopes, with inadequate bandwidth would not yield correct indications of narrow spikes. Therefore, it behooves the practitioners to remain alert to technical facts of life and to make his own interpretations.

The Dangers of Accepting Measurements at Face Value

Many of the voltage measurements made in ordinary electronics pertain to dc levels and to sine waves. It is only too easy to fall into the habit of quickly grabbing any measuring instrument available in order to make these measurements. Fortunately, dire results from such spontaneous action seem to be relatively infrequent. In solid-state power electronics, however, it is prudent to exercise more caution in arriving at conclusions from measurements. Otherwise, catastrophic destruction may result. At best, undesirable performance and unrewarding analysis is all too likely.

While it is always necessary to select meters in terms of their sensitivity, impedance, and maximum range, it is often forgotten that the meter must also be related to the *waveform* being evaluated. In this respect, meters have three important features, one being the waveform they inherently respond to, and another being the kind of waveform values their *scales* are calibrated to indicate. For example, a meter may actually detect the peak-to-peak excursions of a waveform, but the scale may be calibrated to provide readout of rms values. But *even this information* may not suffice to obtain accurate readings, for, thirdly, we must also know for what type of waveform the scale calibration was made. If, as is often the case, the rms scale calibration was made for sine waves, the meter may yield erroneous readings when attempts are made to obtain the rms value of a low-duty-cycle rectangular waveform.

Considering that such measurement values as peak-to-peak, rms, and average are geometrically related parameters, one cannot go wrong by using the *oscilloscope* as a voltmeter, and making the necessary computations to convert the displayed peak or peak-to-peak values to rms or average values. However, a meter is often more convenient to use and is capable of high accuracy if we know the nature of the waveform and the behavior of the meter. Table 1-2 depicts the situations prevailing for various meters and for commonly encountered waveforms in solid-state power electronics. Wherever we see the factor of unity (1.000), the meter readout is *true*. Other factors are multipliers needed to correct the observed reading to true value. Interestingly, some combinations of meter types and waveforms produce *no* true readings for any of the four geometrical parameters!

Thyristor Waveforms

The waveforms shown in Fig. 1-2 are those associated with a simple triac phase-control circuit and offer additional opportunities for erroneous interpretation of measurements. If one studies the rms value of the currents, it is obvious that the meter used for making the measurements must be responsive to the rms value of very nonsinusoidal waves. Thermocouple meters, specially designed electronic meters, and some digital types, can satisfy this requirement most readily. The iron-vane meter and the dynamometer type may also qualify, but both of these meters tend to depart from accuracy with certain nonsinusoidal inputs. It is always best to learn about the response of a meter, rather than assume it is immune to the effects of waveform distortion. (For example, the fixed and movable coils of the dynamometer type sometimes provide sufficiently different reactance to the harmonics of a nonsinusoid input to result in an inaccurate indication.)

Further investigation of the data presented in Fig. 1-2 reveals features contrary to "common sense." For example, note that the power factor of the load is unity in all instances as one would expect from a purely resistive load. It is natural to suppose that power factors less than unity are *only* the result of phase displacement between voltage and current because of inductance or capacitance. Yet, we see that the *line* power factor rapidly departs from unity as the conduction waveform is reduced from 180 degrees. Inasmuch as we are dealing with a simple series circuit, it may appear strange that the power factor is *different* on the two sides of the triac. There is a difference between these two circuit sections, however. Specifically, on the line side, the distorted current wave is compared to the *sine* wave of voltage from the line; on the load side, the voltage and current waves are always *identical*. On the line side, only the *fundamental* of the current wave can be said to be in phase with the line voltage. Many harmonics of the current wave are not in phase with the line voltage—it is as if these phase differences were caused by a physical reactance.

The practical consequence of such low-power-factor operation is that the circuit is much less efficient at low conduction angles than one would infer from the near-ideal switching behavior of the triac. For a given amount of real power delivered to the load, the electric utility must supply more current than would be necessary at unity power factor on the line.

Although meters tend to be more convenient than oscilloscopes, it is obvious that the oscilloscope is indispensable as a means of qualitatively ascertaining the nature of the waveform being measured by a meter. With regard to digital meters, many types have been marketed. There now appears to be a trend toward rms-responding types, and some of these have the desirable feature that the response is accurate for a wide variety of waveforms. One thing to keep in mind when using these instruments is their input impedance, which in some instances is low enough to upset certain circuits. Whether digital or analog, a meter that is capable of providing an accurate readout *regardless* of input waveform is exceedingly useful for the measurements encountered in solid-state power electronics.

(A) Circuit.

Fig. 1-2. Simple triac phase-control

1	2	3	4	5	6	7	8
LINE VOLTAGE WAVEFORM (SINE WAVES)	CURRENT CONDUCTION ANGLE IN DEGREES	LINE AND LOAD CURRENT WAVEFORM. (LOAD VOLTAGE WAVEFORM.)	EFFECTIVE CURRENT (RMS VALUE)	EFFECTIVE VOLT-AMPERES	ACTUAL POWER IN LOAD RESISTANCE	POWER FACTOR AT AC LINE.	POWER FACTOR AT LOAD
$Em1$	180	$Im1$	0.707 Im	0.500 $EmIm$	0.500 $EmIm$	1.000	1.000
	150		0.697 Im	0.493 $EmIm$	0.486 $EmIm$	0.985	1.000
	120		0.634 Im	0.448 $EmIm$	0.402 $EmIm$	0.897	1.000
	90		0.500 Im	0.354 $EmIm$	0.250 $EmIm$	0.707	1.000
	60		0.313 Im	0.221 $EmIm$	0.098 $EmIm$	0.443	1.000
	30		0.120 Im	0.085 $EmIm$	0.014 $EmIm$	0.170	1.000

(B) Waveforms.

circuit and associated waveforms.

15

Table 1-2. Waveform Conversion Factors for Various Types of Voltmeters

WAVEFORM		Peak-to-Peak Responding (OSCILLOSCOPE ELECTRONIC-METER Peak-to-Peak Scale)	Peak Responding (ELECTRONIC-METER Peak Scale)	Peak Responding (ELECTRONIC-METER RMS Scale (Calibrated for sine waves) Also some DIGITAL METERS (not common))	RMS Responding (IRON VANE THERMOCOUPLE DYNAMOMETER RMS ELECTRONIC-METER RMS Scales (Calibrated for sine waves) Also some DIGITAL METERS (newer types))	Average Responding (RECTIFIER-METER RMS Scale (Calibrated for sine waves) Also most older DIGITAL METERS)	Average Responding (RECTIFIER-METER Average Scale (Calibrated for sine waves))
SINE:	pk-pk	1.000	2.000	2.828	2.828	2.828	3.140
	0 pk	0.500	1.000	1.414	1.414	1.414	1.570
	rms	0.353	0.707	1.000	1.000	1.000	1.111
	avg	0.318	0.637	0.900	0.900	0.900	1.000
RECTIFIED SINE: (FULL WAVE)	pk-pk	1.000	1.000	1.414	1.414	1.414	1.570
	0 pk	1.000	1.000	1.414	1.414	1.414	1.570
	rms	0.707	0.707	1.000	1.000	1.000	1.111
	avg	0.637	0.637	0.900	0.900	0.900	1.000
RECTIFIED SINE: (HALF WAVE)	pk-pk	1.000	1.000	1.414	2.000	2.828	3.140
	0 pk	1.000	1.000	1.414	2.000	2.828	3.140
	rms	0.500	0.500	0.707	1.000	1.414	1.570
	avg	0.318	0.318	0.450	0.637	0.900	1.000

		Col 1	Col 2	Col 3	Col 4	Col 5	Col 6
SQUARE	pk-pk	1.000	2.000	2.828	2.000	1.800	2.000
	0 pk	0.500	1.000	1.414	1.000	0.900	1.000
	rms	0.500	1.000	1.414	1.000	0.900	1.000
	avg	0.500	1.000	1.414	1.000	0.900	1.000
RECTIFIED SQUARE	pk-pk	1.000	1.000	1.414	1.414	1.800	2.000
	0 pk	1.000	1.000	1.414	1.414	1.800	2.000
	rms	0.707	0.707	1.000	1.000	1.272	1.414
	avg	0.500	0.500	0.707	0.707	0.900	1.000
RECTANGULAR PULSE	pk-pk	1.000	1.000	1.414	$1/D^{1/2}$	$0.9/D$	$1/D$
	0 pk	1.000	1.000	1.414	$1/D^{1/2}$	$0.9/D$	$1/D$
	rms	$D^{1/2}$	$D^{1/2}$	$1.414\,D^{1/2}$	1.000	$0.9/D^{1/2}$	$1/D^{1/2}$
	avg	D	D	$1.414\,D$	$D^{1/2}$	$0.9\,D$	1.000
TRIANGLE	pk-pk	1.000	2.000	2.828	3.464	3.600	4.000
	0 pk	0.500	1.000	1.414	1.732	1.800	2.000
	rms	0.289	0.577	0.816	1.000	1.038	1.153
	avg	0.250	0.500	0.707	0.867	0.900	1.000

Also appearing on the market are various "smart" oscilloscopes. These instruments are associated with calculator ICs or microprocessors in such a way that a digital readout of the various wave values is quickly and conveniently presented in addition to the conventional waveform patterns. Also, rates of rise and decay are speedily and accurately indicated. It is anticipated that these instruments will prove very useful in solid-state power electronics.

In electrical power work, it is customary to use wattmeters and power-factor meters to evaluate the performance of motors, heaters, and other loads. However, when the rms current to a load is controlled by a thyristor, the use of these instruments becomes suspect. Again, it is wise to establish the response characteristics of such meters before making measurements in SCR or triac circuits. Often, corrective curves can be plotted so that departures from accuracy can be remedied as readings are recorded, or afterwards.

The examples given for triac operation also apply to back-to-back operation of SCRs. At the present, SCRs have greater power-handling capability than do triacs. Whereas, triacs may be found in light dimmers and in control circuits for hand tools, SCRs will be encountered in applications involving large industrial motors, or in traction vehicles. Single SCR circuits behave similarly, except that only half-wave power is delivered to the load. Here the chances of grossly erroneous meter readings are even greater, for the current wave is mathmatically nonsinusoidal even at current conduction angles of 180 degrees.

THE BASIC TECHNIQUES FOR
MANIPULATING ELECTRICAL POWER

Our investigation of solid-state power electronics will be considerably aided if we first consider the ways in which solid-state devices may be used to process or control electrical power. The number of different operating modes that have come into use will likely prove surprising. It is significant to note that, although some of these techniques represent long-known concepts, they have only recently been implemented in practice.

On-Off Switching

Nothing is implied here beyond the mere turning on and off the power applied to a load. Either dc or ac may be applied and interrupted. A switch in the ac power line may be used, the solid-state power device may be connected and disconnected from the power source, or the trigger or control electrode of the power device may be actuated to render the device conductive or nonconductive. This appears simple enough from one's experience with low-power equipment. But in every circuit, there is at least *stray* inductance and capacitance. When current is made or interrupted, a "stray" emf will be generated according to the relationship, $e = L \, di/dt$. Also, an "unintended" current, i, will flow according to the relationship, $i = C \, dv/dt$. These switch-produced voltages and currents are known as *transients* or *surges*. When turning on and off large blocks of power, it is usually *not* sufficient to merely arrange for making or breaking contacts, or for abruptly making a solid-state device conductive or nonconductive. For, even a one-time transient can be in the nature of a destructive arc, or a voltage spike greatly exceeding the "dielectric" strength of a control device, or a current surge exceeding the maximum current rating of a device.

Somewhat more insight is provided by contemplation of the energy equations which are operative at all times in electrical circuits. They are:

$$W = \frac{I^2 L}{2}$$

or,

$$W = \frac{E^2}{2C}$$

where,
  W is in watt-seconds or joules,
  I is in amperes,
  E is in volts,
  L is the inductance in henrys,
  C is the capacitance in farads.

These equations tell us that, even though L and C may exist primarily due to wiring and the proximity of circuit board components, the stored energy may be appreciable if E or I are high. For example, the energy stored in the magnetic field of a given conductor is one million times greater if the steady-state current is ten amperes, rather than ten milliamperes.

It happens that there is not much we can do about the stray energy stored in the magnetic and electric fields of a power circuit, once we have arranged for a reasonable wiring layout. And to make matters worse, energy storage will often be much higher in the *physical* capacitors and inductors in most circuits. For example, if a simple rectifier-filter system is turned on, the filter capacitor initially appears as a short circuit to the applied dc. As the capacitor charges, this condition is alleviated. However, at the first instant the rectifier elements are subjected to a tremendous current surge. In power equipment, such a surge can easily destroy these rectifiers. At best, the designer may be forced to provide expensive rectifiers which have continuous operation ratings greatly in excess to that needed.

Although we have been discussing the simple *one-time* turn on and turn off of power equipment, this discussion also applies to numerous techniques wherein power is switched periodically, such as inverters, switching-type regulators, and phase-controlled thyristor circuits. And, because of energy storage and switching transients, almost all solid-state power circuits make use of *suppression* techniques. That is, the energy content of the transients are either reduced or their energy is largely absorbed in some device designed for this purpose. One way to reduce the energy content of switching transients is to slow down the rise and fall times of the switching process. Unfortunately, although this is readily accomplished with the one-time turn on or turn off of equipment, it can only be applied in a limited way to any circuit involving repetitive switching. In such circuits, the efficiency of the switch suffers if too much time is consumed in turn on and turn off. (Any finite time spent by a switch in getting from one state to another is a time of power dissipation in the switch itself.) Any practical design must balance these conflicting needs for rapid and limited-rate transition times in a switching device.

Another reason for suppressing switching transients is that they are a source of *interference* to other circuits and systems. Much of the energy present in sharp switching transients radiates and is particularly difficult to combat once it finds its way into sensitive circuitry, such as logic systems. So, do not be surprised to find common transient-suppression techniques used in otherwise diverse solid-state power circuits.

Power Control Via Switching Techniques

It is surprising how many different ways there are to control the amount of electrical power consumed by a load. If the power source is dc, the most obvious control element is some form of rheostat or potentiometer. This is entirely acceptable and has much to commend it. For example, such control is easily made smooth, continuous, wide range, and may readily be made proportional to a quantity such as shaft rotation or programming voltage. However, one must be reconciled to live with the "throwaway" power that must be dissipated in the control element. Unfortunately, in power electronics, it often proves neither economic or otherwise permissible to operate at the inherently low efficiencies provided by such resistive control. Appropriately enough, such a control technique, whether implemented by a physical rheostat or the latest monolithic Darlington transistor, is known as *dissipative* control. Because of the problems of cost and heat removal associated with dissipative control, more efficient power-control methods are used in power electronics.

With regard to the control of dc power, it often proves profitable to convert it to ac which is subject to more efficient control techniques, then convert back to dc via rectification and filtering. Another much-used technique is to "chop" the dc and pass the rough waveform through a low-pass filter before presentation to the load. The power delivered to the load can be controlled by means of either of two chopping or switching modes. The pulsed dc can be duty-cycle modulated with the chopping frequency remaining constant. This is generally referred to as *pulse-width modulation*. Or, the pulse width can be maintained constant while the frequency or repetition rate is varied. In both cases, the average voltage, Vo, presented to the load is given by the equation:

$$V_o = \left(\frac{T_{on} - T_{off}}{T_{on}} \right) V_{in}$$

where,
    V_o is the average voltage presented to the load,
    T_{on} is the on time,
    T_{off} is the off time,
    V_{in} is the source voltage.

The waveforms pertaining to these control methods are illustrated in Fig. 1-3. The low-pass filter can be thought of as an averaging or integrating circuit in addition to an attenuator of the switching-frequency pulses and its harmonics. The freewheeling diode enables load current to be supplied during T_{off}. The source of this current is the energy stored in the magnetic field of the inductor.

The beauty of the *ideal* switching device is that it dissipates *no* power when in the ON state because it displays zero resistance; likewise, it dissipates *no* power in the OFF state because its resistance is infinite. Actual solid-state switching devices do not yield such 100 percent control efficiency. They exhibit resistance when ON and a small, but significant, leakage current may flow when they are turned OFF. Moreover, these devices also have dissipative losses because they can neither turn ON or OFF in zero time. Although their rise and fall times may be small, the losses can be appreciable if these transition times are comparable to the ON time. Nonetheless, power control by these switching techniques is capable of much greater efficiency than can be attained with dissipative control.

A more detailed view of the switching waveforms is illustrated in Fig. 1-4. Here, we see that the ideal "pure" dc is not achieved for V_o. Rather, there is a superimposed ripple component. The operation of the inductor is also shown in Fig. 1-4. Note that the inductor voltage makes positive and negative excursions around the average value, V_L. The average dc value of the inductor voltage is zero. The inductor current is essentially the load current with a superimposed sawtooth. Because of the freewheeling diode, the inductor delivers load current

CONSTANT FREQUENCY,
VARIABLE WIDTH

CONSTANT WIDTH,
VARIABLE FREQUENCY

Fig. 1-3. Switching waveforms in dc power-control systems.

not only when the switch is closed, but also when it is open. These switching waveforms pertain to both the variable width and the variable frequency modes of operation. The output ripple voltage, although a smooth wave, is not sinusoidal. In practical dc switching supplies, it is not easy to reduce the output ripple to an extremely small value because the electrolytic filter capacitors have appreciable equivalent series resistance. And increasing the size of the inductor adversely affects speed of response in a regulated supply, to say nothing of cost, weight, and packaging problems.

A particularly advantageous scheme of dc power control is to rectify and filter the incoming ac from the power line, control the resulting dc by means of a switching device, and then pass the chopped waveform through a low-pass filter to the load. It often happens that a 60-hertz power transformer is not needed because this type of switching control can operate efficiently at a low duty cycle in order to deliver a dc load voltage considerably below line voltage. (If this is attempted with a dissipative control, such as the common series-pass transistor in regulated power supplies, the power lost in the rheostat-like control device may easily exceed the power delivered to the load.)

A somewhat more sophisticated version of the above system uses the rectified and filtered ac line power to operate a dc to ac inverter which has its ac duty-cycle modulated or programmed. Of course, the ac from the inverter must then be rectified and filtered again so that

Fig. 1-4. A more detailed view of switching waveforms.

the load may have controlled dc power applied to it. At first it might appear that this is a roundabout approach that unnecessarily requires extra components. Actually, the savings in size, weight, and cost that can be effected by this dc power control method are considerable. This comes about by virtue of the *high* operating frequency of the inverter. This operating frequency may be from 1 to 50 kilohertz, or even higher. A good balance among many contradictory factors involving various frequency-dependent losses, together with the costs of components, dictates a switching rate in the vicinity of 20 to 25 kilohertz. This is above the general audible threshold, sometimes an important consideration in power-control systems because transformers and chokes can radiate annoying levels of accoustic energy. These relatively high frequencies require only minuscule core components. The elimination of 60-hertz magnetics is then the compelling feature of such a system—the few extra parts are trivial contrasted to the gains achieved by dispensing with the *bulky* and expensive 60-hertz transformer.

Most dc power-control systems, whether they utilize dissipative or switching techniques, are generally closed-loop arrangements which provide either constant voltage or constant current to the load regardless of wide variations in ac line voltage, or in the load itself. Such automatic stabilizing systems are collectively known as regulated power supplies. They constitute one of the most important types of circuits encountered in solid-state power electronics.

AC Power Control—The Phase Control Method

It is obvious that ac power can be controlled by rheostat-like devices in similar fashion to the dissipative control of dc power. And likewise, one must then be prepared to cope with low efficiency, high cost, and awkward thermal problems. In principle, ac power may be controlled with high efficiency and low dissipation by means of variable reactance. Practical implementations have employed such devices as magnetic amplifiers and saturable chokes. Thus far, no practical utilization of variable capacitance has emerged because the physical size of such a device at power-line to supersonic frequencies would rule out its consideration.

One of the most popular power-control methods is the phase control of the ac sine wave by means of thyristors. Fig. 1-5 illustrates both half-wave and full-wave voltage waveforms in which the rms power delivered to the load is governed by the timing of the gate turn-on trigger applied to the thyristor. The reason underlying the high efficiency of this type of control is that its operation is essentially that of a *switch*—it is either fully ON or completely OFF (ideally). The half-wave voltage shown in Fig. 1-5A is typical of single SCR control of the load. The full-wave scheme (Fig. 1-5B) can be implemented either by a single triac or by back-to-back SCRs. SCRs with power-handling capabilities greatly exceeding other solid-state devices have become available at very nominal cost. Also, the voltage and current capabilities of triacs are constantly being improved. These considerations tend to override the disadvantages inherent in the very non-sinusoidal waveforms produced. These waveforms produce electrical noise, are difficult to measure with commonly available instrumentation, and can result in operation at low power factor.

(A) Half-wave voltage waveform.

θ = ANGLE OF CONDUCTION

(B) Full-wave voltage waveform.

Fig. 1-5. Load-voltage waveforms in phase-controlled power systems.

Pulse-Modulation Method of AC Power Control

The load voltage waveforms associated with two pulse-modulation techniques are shown in Fig. 1-6. These waveforms are usually produced by inverters. In Fig. 1-6A equal-duration pulses of varying amplitudes are appropriately timed to approximate the general contour of a sine wave. If each of the wave "building blocks" is made proportionately greater or smaller in amplitude, the rms value of the composite wave is varied, but its essential shape remains unaffected. If such a waveform is "strained" through a low-pass filter, a reasonably good sine wave may result. In practice, the windings of a motor may suffice for this purpose.

The simulated sine wave depicted in Fig. 1-6B is pulse-width modulated—the amplitude remains constant, but the width of the pulse is apportioned so that the distribution of power within the composite wave is approximately sinusoidal. Power control may be affected by varying the widths of the pulses, but at the same time maintaining their *relative* durations. As with the waveform in Fig. 1-6A, a low-pass filter may be used to produce an acceptable sine wave.

These "fabricated" ac sine waves are often found in power-control systems in which the control parameter is frequency, rather than rms value. Special considerations then apply to the widths of the pulses in both waveforms. However, this is easily taken care of in low-power logic circuitry. These wave-construction techniques also are useful in designing three-phase power equipment. Again, the phase displacements between the waveforms are accomplished in low-level logic

(A) Pulse amplitude modulation.

(B) Pulse width modulation.

Fig. 1-6. Load-voltage waveforms in pulse-controlled power systems.

circuits. Required flexibilities in control functions are more readily attained with these fabricated waveforms than by resorting to "brute-force" analog methods. For example, a speed-control system for three-phase induction motors requires that the applied motor voltages go up and down with the impressed frequency. In addition, there are power-output circuits configured about both transistors and thyristors. (Although their power-handling abilities are relatively limited, both *gate turn-off SCRs* and *power MOSFETS* merit consideration for such service.)

Burst-Modulation Method of AC Power Control

It had long been apparent that for certain types of loads, such as heating elements, it was possible to control average ac power in yet another way. If varying lengths of ac wavetrains are metered out to a load, an *average* power level will be dissipated in the load which will be less than would be experienced if the ac cycles were continuously applied. And, the greater the length of time between successive wavetrains, the less will be the average load power. However, for this basic technique to achieve maximal usefulness, a certain refinement is desirable. Specifically, only complete sine waves should be metered out to the load. Otherwise, one can anticipate considerable rfi and emi. Any time a sine wave is chopped so that there are fractional cycles between voltage zero crossings, the wavetrains are nonsinusoidal from a methematical viewpoint. This, alone, can result in the production of many higher harmonics capable of creating interference. Additionally, the application of voltage at nonzero levels is tantamount to high dv/dt rates, which further enriches that portion of the electromagnetic spectrum contributing to electrical noise.

If, on the other hand, this control technique is implemented under the condition that only wavetrains comprising *complete cycles* are metered out to the load, there will be no rfi, emi, or harmonics produced. Fig. 1-7 illustrates such power control. Note that these load voltage waveforms commence and cease only at *zero crossings*. Of course this control scheme is best adapted to "sluggish" loads—those which would not be likely to respond with intermittent operation. This restricts its use mainly to heaters, but under some circumstances motor speed control is also feasible.

EVALUATING POWER AMPLIFIERS

There are many solid-state devices which are used as power amplifiers, and there are quite a few circuit techniques available for accomplishing power amplification. Accordingly, it would be useful if the basic performance of power amplifiers could be described by a standardized figure of merit. That is, we should be able to visualize any device and/or any circuit technique capable of developing amplification at high power levels as a "black box" possessing a meaningful quality as a power-level booster. In the technical literature, we find that transistors are commonly specified in terms of their current gain at ordinary frequencies and by their power gain in the uhf and microwave portion of the spectrum. However, the amplifying ability of

(A) Maximum load power.

(B) High load power.

(C) Low load power.

(D) Zero load power.

Fig. 1-7. Load-voltage waveforms in burst-modulation-controlled power systems.

power FETs is mainly specified by their transconductance. When either device is used in a pulse-width-modulated, or Class-D, amplifier, there seems to be no set rule for relating the output to the input.

Actually, all amplifiers are *power* amplifiers even though we often term them otherwise, such as voltage or current amplifiers. (If we are truly interested in voltage or current gain *only,* the ordinary *transformer* can provide such "amplification.") An appropriate figure of merit must involve *both* voltage and current in such a way that one isn't simply gained at the expense of the other such as in a transformer. The use of the term power gain involves practical difficulties for such a device as the power FET in which input power is negligibly small. Similarly, the power FET is not readily described by current gain because of its minute input current. This current would be difficult to measure, and it would not be easy to determine what portion of it was due to leakage effects. Voltage gain of the power FET might lead to some useful results, but would fall short of practicability when applied to bipolar transistors. In any event, voltage or current gains must be related to input and output impedances to be truly meaningful for power amplifiers.

It turns out that the concept of *transconductance* in which output current is related to input voltage is applicable to *any* black box serving the function of amplification. See Fig. 1-8. Not only is this figure of merit used with the new power FETs, but manufacturers of bipolar power transistors are providing transconductance data in their specifications to an increasing extent. The transconductance of any device or circuit is given as the ratio of a small change in *output current* to the change in *input voltage* causing the change in output current, while all other operating parameters are maintained constant. Interestingly, whereas anything in excess of 10,000 micromhos was considered a "hot" vacuum tube, power transistors can have transconductance values in the *millions* of micromhos range. From such curves as

$$\text{TRANSCONDUCTANCE} = \frac{\text{SMALL CHANGE IN OUTPUT CURRENT}}{\text{CAUSATIVE CHANGE IN INPUT VOLTAGE}} = \frac{\Delta I_{OUT}}{\Delta V_{IN}}$$

Fig. 1-8. Transconductance concept suitable for solid-state power amplifiers.

are shown in Fig. 1-9, the transconductance at an operating point such as "P" is determined by the equation:

$$gm = \frac{\Delta I_{out}}{\Delta V_{in}}$$

where,
gm is the transconductance,
ΔI_{out} is a small change in output current,
ΔV_{in} is a small change in input voltage.

Note that the collector* voltage must be held constant when applying this concept.

Actually, many designers of transistor circuitry have continued to use transconductance as a heritage from their vacuum-tube days. The transconductance of a transistor can be derived from the relationship, $gm = h_{fe}/h_{ie}$. However, it is more conveniently determined from graphs of collector current vs. base-emitter voltage which power transistor manufacturers are now providing with their specification sheets. Although the true mathematical use of transconductance precludes the insertion of a load in the output circuit, many amplifier circuits used in power electronics tend to have about the same transconductance as

---

* In vacuum tubes, the plate (and screen grid, if present) must be held at a constant voltage. When using FETs, the drain voltage must be held constant.

Fig. 1-9. Transconductance characteristics of the Delco DTS-424 power transistor.

do the active devices alone. This is because these amplifiers are for the most part not employed as so-called "voltage amplifiers," since the loads have relatively low output impedance. Yet another reason why the approximation is usually good is that power transistors and power FETs display "pentode" characteristics. That is, they behave as constant-current generators (output current is about the same with a zero-impedance load or the actual load).

While the ability to visualize any amplifier as a "transconductance stage" provides operational insight, one must pay heed to other factors effecting the practical implementation of a power amplifier. Consider, for example, two Class-A power amplifiers displaying the same overall transconductance, but one is designed with bipolar transistors, whereas the other uses FETs. A significant practical difference is that the FET amplifier will consume negligible current from the signal source. Its bipolar transistor counterpart, on the other hand, must be supplied with substantial signal current. (Although it is easy to say that the FET amplifier has the higher "power gain," this does not argue well for the use of power gain as the standard yardstick for evaluating amplifiers. A little reflection reveals that *any* FET device may be assigned infinite power gain when operating as a Class-A amplifier.) Also, transconductance often must be used with other relevant parameters such as input capacitance, frequency limits, voltage and current boundaries, etc. But, transconductance can be a good place to start in the design or the analysis of a solid-state power amplifier. The transconductance concept can be directly applied to other types of solid-state power amplifiers, such as parametric amplifiers, negative impedance amplifiers, Class-D amplifiers, and to magnetic amplifiers using diodes.

CLASS-A AMPLIFIER

The Class-A amplifier delivers a proportional output over the entire cycle of input signal. In other words, this is a linear amplifier. It is not surprising that the Class-A power amplifier was extensively used in audio equipment. It is not very popular any longer for higher power audio amplifiers because of its inherently poor efficiency, which approaches 50 percent under ideal conditions. Inasmuch as audio signals have high peak-to-average ratios, the efficiency of the Class-A amplifier is quite low in actual sound applications. For a long time this was tolerated because of its excellent linearity. In particular, push-pull Class-A power amplifiers tended to cancel even-harmonic distortion. Such audio equipment often produced high-quality reproduction with minimal local or loop feedback. Indeed, there remains a number of devout stereo buffs who feel that Class-A power amplification is the superior mode for faithful audio reproduction. However, when the consumer demand increased for high power capability, Class-A amplification had to give way to more efficient methods. Although many users would have been willing to cope with the higher cost of operation of Class-A audio power equipment, the thermal problems proved incompatible with the needs of consumer products. This can be appreciated in light of the fact that a Class-A amplifier dissipates half of its available sinusoidal output power while just idling with no input signal.

Class-A amplification is still extensively used at lower power levels, such as the driver stages in audio circuitry. Class-A amplifiers are often found in industrial applications where efficiency may not be a paramount consideration, and where ample space and cooling apparatus can readily be provided. The series-pass device in linear regulators, while not commonly referred to as a Class-A amplifier, nonetheless operates in a fashion resembling this amplification mode. As with more conventional Class-A power amplifiers, the series-pass device is inefficient because it must dissipate considerable power within itself.

Single-Ended Class-A Amplifier

Typical single-ended Class-A power amplifiers are shown in Fig. 1-10. The dynamic operation of these amplifiers is illustrated in Fig. 1-11. Note that the operating point is established to enable optimum utilization of the linear portion of the transfer characteristic curve. The operating point is determined by all three of the biasing resistances, R1, R2, and R3. Unless adequately by-passed over the operating frequency range, emitter resistance R1 also produces current-derived negative feedback. Although such feedback lowers the stage gain, it reduces certain distortion products. Also, the input and output impedances of the stage can often be conveniently manipulated by such feedback. This resistance is very beneficial in stabilizing the operating point against temperature variations, particularly with germanium transistors where it can help prevent thermal runaway.

The Push-Pull Mode for Class-A Operation

A typical push-pull power amplifier circuit suitable for Class-A operation is shown in Fig. 1-12. The operating characteristics of such a circuit is illustrated in Fig. 1-13. Note that the operating points of transistors Q1 and Q2 are so situated that the output-current swing involves *more* than just the linear portion of the transfer curves. This depicts one of the compelling features of push-pull operation. Even though each transistor produces considerable second-harmonic distortion, the *composite* output waveform is free of such distortion. This implies that for a given amount of output distortion the push-pull arrangement can yield *more than twice* the output power practically attainable from a single-ended Class-A amplifier. Inasmuch as low operating efficiency is the principal shortcoming of Class-A amplifica-

(A) Optimum power gain by use of input impedance-matching transformer.

(B) Simpler input coupling arrangement.

Fig. 1-10. Typical Class-A power amplifiers.

tion, any expedient which allows greater output to be developed is certainly welcome.

In order to provide clearer insight into the nature of second-harmonic cancellation, the second-harmonic current waveforms in the two halves of the output transformer primary winding are depicted in Fig. 1-13. These second-harmonic components would *not* be observed by probing with an oscilloscope. Only the badly distorted output current waveforms of Q1 and Q2 would be seen on the scope. (However, either a mathematical or a graphical analysis of these current waveforms would yield their constituent harmonic "building blocks." Briefly, the *Fourier Theorem* states that any nonsinusoidal wave can be resolved into a number of harmonically related sine waves. The basic repetition rate of the nonsinusoidal wave is the fundamental, or "first harmonic," of the Fourier series of constituent frequencies. Often, there is also a zero-frequency or dc component in the series, particularly when *even* harmonics are present. The phase relationships and amplitudes of the various harmonics give rise to any imaginable waveshape.)

32

Fig. 1-11. Representative operating characteristics of all Class-A power amplifiers.

In our particular case, it can be seen that the second-harmonic waves *cancel* because of their equal amplitude and 180-degree phase relationship. The algebraic addition of the Q1 and Q2 currents take place in the secondary winding of the output transformer. The net

Fig. 1-12. A typical push-pull power amplifier circuit.

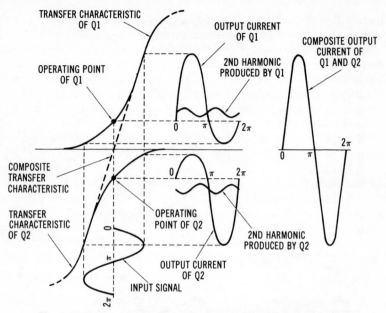

Fig. 1-13. Second-harmonic cancellation in hard-driven Class-A push-pull amplifier.

result, as seen by the load, is a nice clean sine wave—or at least one devoid of obvious second-harmonic distortion. It should be pointed out that other even-order harmonics are generated by the curvature of the transfer curves of Q1 and Q2. These even-order harmonics, such as the fourth and sixth harmonics, are *also* cancelled in similar fashion to the predominant second harmonic. It should be added that transfer curve nonlinearity can also produce odd-order harmonics, the third, fifth, seventh, etc. These odd harmonics distort the smooth sinusoidal shape of the composite output current waveform, but tend to produce less troublesome effects than do even harmonics, unless present in unusually high amplitudes.

There are even other advantages of Class-A push-pull operation. There is no dc saturation of the output transformer because the center-tap connection causes the direct currents in the two halves of the primary winding to produce opposing fluxes. Not only does this make for a more economical transformer design, but a potential source of distortion products, principally odd harmonics, is eliminated. Bypass capacitors are less critical in the push-pull circuit because signal components need not complete their paths through the dc power supply. And, the nature of push-pull operation is such that less imposing demands are made on power-supply filtering. This is because ripple fed to the center-tap of the output transformer tends to cancel. However, there will not be any discrimination against power-supply ripple mixed with, or superimposed upon, the input signal.

It should be mentioned that the advantages of push-pull operation deteriorate rapidly when the two active devices are not identical. It has not always been easy to obtain pairs of power transistors with reasonably matched characteristics. However, if one ascertains that the current gain factors (beta) are matched for corresponding currents and frequencies, good results should be obtained. It is generally not sufficient to merely match current gains at dc or at a very low frequency. A wise strategem in power amplifier design is to attend *first* to those matters pertaining to the best linearity and balance that can be had; *then* sufficient negative feedback may be incorporated to further minimize signal distortion. The notion of such feedback as a universal panacea for sloppy amplifier design is a cosmetic approach which never completely disguises the underlying distortion. Many a stereo amplifier fails to convey acoustical integrity to the ear despite decades of flat frequency response, tremendous dynamic power capability, and, of course, multiple decibels of feedback.

CLASS-B AMPLIFIER

The Class-B amplifier delivers output current and power only over one-half cycle (180 degrees) of an applied sine-wave input signal. Inasmuch as such operation is suggestive of rectification, it is obvious that a single-ended Class-B device would produce extremely high distortion in audio applications. However, the push-pull configuration (Fig. 1-14) solves this problem. This is because only one device is conductive at any given time—the other device is cut off until polarity reversal of the input signal alternates the conductive state of the two devices. By the same token, when no input signal is applied, the idling current of the push-pull Class-B amplifier is zero. As a result, one would anticipate higher operating efficiency than is available from Class-A amplifiers. It turns out that the Class-B amplifier has a theoretical maximum efficiency of 78 percent. Such efficiency prevails only when a Class-B output stage is driven to its full rated output. Therefore, the operating efficiency with audio signals is generally much lower. Nonetheless, the average efficiency tends to be appreciably greater than is forthcoming from the Class-A amplifier operating under similar signal and output conditions. And this is despite the fact that the input-power requirements of most devices operated in Class B is greater than for Class-A operation.

Class-B power amplification is commonly used in servo amplifiers, and in other "brute-force" tasks devoted to power gain. Paradoxically, Class B is not the predominant mode encountered in audio work—at least, not in its "pure" form. This is because most devices are not

Fig. 1-14. Power devices operating essentially in the Class-B mode.

linear at low-power operation levels which gives rise to a switching transition known as *crossover distortion*. The best way to avoid the most detrimental effects of crossover distortion is to blend in a small amount of Class-A operation. That is, although the power amplifier is essentially Class B in the sense that its average dc current swings with the signal, its quiescent output current is relatively low so that the high average dissipation which characterizes Class-A operation does not exist. At the same time, the switching transition from one of the push-pull devices to the other is smoothed out and crossover distortion is greatly reduced. The significant feature of the Class-B push-pull amplifier is that the full waveform is reproduced in the output circuit even though each device is active for only one half of the input wave.

COMPLEMENTARY-SYMMETRY AMPLIFIERS

The concept of complementary symmetry has been applied ever since both npn and pnp transistors became available. Complementary symmetry allows simple push-pull amplifier stages to be designed. Inasmuch as there is no vacuum-tube version of the pnp transistor available, this interesting circuit certainly could not have been implemented prior to semiconductor technology. As a matter of fact, until recently, the serious deployment of the technique was hampered by the lack of suitable pnp power transistors to form matched pairs with high-quality silicon npn types. Not only has this situation been remedied, but complementary pairs of power FET devices are also being introduced. The technique of complementary symmetry is applicable to Class-A, -B, -C, -D, -G, and -H amplifiers. At the same time, it simplifies bridge circuits, as well as push-pull configurations.

Fig. 1-15 shows the basic concept of complementary symmetry. Note that the bases of the power transistors are connected together. This might not initially appear conducive to push-pull operation, but a little consideration will reveal that each transistor is made conductive by its own half cycle of an ac input signal. Also, Class-B operation automatically ensues. Actually, in order to eliminate the dead zone (where neither transistor is conductive) and to reduce crossover dis-

Fig. 1-15. Basic complementary-symmetry amplifier.

tortion, both transistors are generally provided with a slight amount of forward bias.

An actual audio-amplifier with complementary-symmetry output is shown in Fig. 1-16. The functions of the important elements involved in the operation of the circuit are as follows: Transistors Q4 and Q5 are a complementary-symmetry pair operating as a Class-AB output stage. (Operation is essentially Class B, with just enough forward bias to practically eliminate crossover distortion.) The fact that the bases of these transistors are not directly connected together stems from their differing bias requirements. The actual application of the signal is such that both bases are essentially driven in parallel. This simple system has but one driver, the differential-voltage-gain stage made up of transistors Q1 and Q2. Providing that an input signal with sufficient amplitude is available, such a signal could be applied to the base of Q1. A resistive network could then be inserted between the junction of R5 and R6 and the base of Q2 in order to provide negative feedback. Transistor Q3 supplies the forward biases to the complementary-symmetry output stage. This could be a ticklish task, for it is known that the base-emitter voltage of each of the output transistors will decrease at the rate of 2 millivolts per

Fig. 1-16. Simplified audio amplifier with complementary-symmetry output stage.

degree Celsius. Besides supplying these transistors with different biases—one being npn, and the other pnp—some automatic tracking arrangement must be incorporated to prevent the output idling current from changing with temperature. Transistor Q3 is able to provide both of these biasing requirements for two reasons. First, it is fed from a constant-current stage, Q6, and second, it is physically mounted on the same heat sink supporting output transistors Q4 and Q5. The constant-current source makes V_{BE} of Q3 dependent upon *temperature* only. The heat-sink mounting technique constitutes thermal feedback which causes Q3 to provide compensatory biases in response to a temperature change. The V_{BE} of Q3 can be thought of as the "reference" voltage in such a thermal/electrical feedback arrangement. Variable resistance R4 adjusts the nominal biases applied to the output transistors.

Suppose that increased temperature tends to cause the quiescent, or idling, current of the output stage to increase. The same temperature increase causes V_{BE} of Q3 to decrease, thereby decreasing the biases applied to the output stage and very nearly restoring the idling current to its original value. Good tracking can occur over a wide temperature range because the physical mechanisms governing output current and applied bias from Q3 are essentially similar—both are predicated upon V_{BE} responses to temperature.

Actually, constant-current source Q6 serves *another* important function. It acts as a high-impedance dynamic load for transistor Q2 in the input stage. This enables the input stage to develop a much higher voltage gain than is readily accomplished with a collector load resistance.

Not only do complementary-symmetry power amplifiers make it easy to dispense with transformers, but it also becomes unnecessary to use a large electrolytic coupling capacitor in the speaker or load circuit when dual power supplies are used.

THE QUASI COMPLEMENTARY-SYMMETRY AMPLIFIER

Basic quasi complementary-symmetry amplifiers are shown in Fig. 1-17. One of the advantages of this type of circuitry over the "pure" complementary-symmetry arrangement is that a pair of identical power transistors are used in the output stage. The reason the quasi circuit came into prominence was because the earlier germanium pnp transistors did not have any germanium npn counterparts. Also, the later silicon npn transistors usually did not have a silicon pnp counterpart. Indeed, it has only been relatively recently that silicon npn-pnp matched pairs have been readily available on the market. It still remains true that there are less silicon pnp transistors available for higher power ratings.

(A) Circuit using npn power output transistors.

(B) Circuit using pnp power output transistors.

Fig. 1-17. Basic circuits of quasi complementary-symmetry amplifiers.

Fortunately, the main features of complementary-symmetry circuitry are attainable from arrangements which make one output transistor behave as though it were of opposite polarity. The basic idea is to use a conventional Darlington connection from a driver on one side of the push-pull circuit and a complementary-symmetry Darlington on the other side. Thus, for the circuits in Fig. 1-17, the upper portion of the circuits (involving transistors Q1 and Q3) comprise *conventional* Darlington combinations of driver and output transistors. Conversely, the lower portion of the circuits (involving transistors Q2 and Q4) utilize the *complementary-symmetry* type of Darlington connection. Insofar as concerns the signal source, power transistor Q4 of the circuit in Fig. 1-17A behaves as if it were a pnp, rather than an npn type. Thus, the output phase relationship between the two halves of the circuit comply with the needs of push-pull operation. In the circuit in Fig. 1-17B power transistor Q4 behaves as if it were actually an npn type.

One might question the degree of balance prevailing when one transistor of the push-pull output pair delivers load current from its emitter, and the other delivers load current from its collector. Surprisingly, it turns out that this is not a major source of imbalance in audio amplifiers, primarily because of the low load impedance presented by speakers. Under such condition, common-emitter and common-collector circuits operate in a very similar fashion. Even so, some manufacturers produce specially designed power transistors intended for optimum performance in quasi complementary-symmetry amplifiers.

The circuits of Fig. 1-17 are simplified and do not include biasing provisions. As depicted, they would produce crossover distortion in audio applications, but would be entirely suitable for certain other uses, such as servo amplifiers. If the circuits shown were used, the relative severity of crossover distortion would be much less if 50- or 60-volt transistors were used than if the amplifier were designed for 12- or 15-volt operation.

CLASS-C AMPLIFIERS

The Class-C amplifier carries the operating concept of Class-B amplification even farther. In a Class-C amplifier, output current flows for only a small portion (120 degrees or less) of an input sine wave. Therefore, two devices operated in Class-C mode in a push-pull arrangement cannot restore the missing part of the input signal, and linear amplification cannot be accomplished. This rules out the use of such an amplifier for reproducing audio signals. The Class-C amplifier finds its principle application in boosting the power level of radio-frequency signals in transmitters. A *resonant circuit* in the output stage provides the means whereby a sustained radio-frequency sine wave is produced for the antenna, or for another Class-C stage. In essence, the resonant circuit is "shock excited" into oscillation. Despite its name, the Class-C amplifier behaves more like a *switch* than an amplifier. If, for example, it were possible to drive an electromechanical relay at a radio-frequency rate, the same type of operation would ensue if the contacts of the relay were associated with a dc power source and a resonant circuit. Think of the Class-C amplifier as a source of current pulses. This being the case, it is only natural for the resonant circuit to oscillate. Although the oscillations so produced are *damped,* they are not perceived to be so if the Q of the resonant circuit is high enough. The *flywheel* effect of such a *tank* circuit then produces an essentially constant radio-frequency output.

Sometimes it is desirable to impart power amplification to a *modulated* radio-frequency signal. If this were attempted with a Class-C amplifier, either a single-ended or a push-pull configuration, the audio modulation would not be reproduced in the output of the resonant tank circuit. Such amplification can, however, be accomplished with amplifier circuits which resemble Class-C stages, but which are biased to operate in the Class-B mode. Such an amplifier is appropriately termed a *linear Class-B stage.* It increases the rf power level by shock exciting the resonant output circuit with current pulses. (Of course, the current pulses are now 180 degrees in "pulse width.") Its operation resembles the switching action of the Class-C amplifier. However, the radio-frequency oscillations excited in the output tank circuit will *follow* the amplitude undulations of the modulated input signal. Either single-ended or push-pull configurations work well in this application.

Class-A amplifiers associated with a resonant output tank circuit can also boost the power level of a modulated rf signal. The chief drawback of Class-A rf power amplification is its relatively low operating efficiency. On the other hand, the generation of radio-frequency harmonics is much lower than with either Class B or Class C. Fig. 1-18 illustrates the basic operational differences in Class-A, -B, and -C amplifiers.

CLASS-D AMPLIFIERS

The development of solid-state power devices has created new interest in techniques that have their roots in the past, but could not be readily exploited during the long reign of the vacuum tube. A Class-D amplifier exemplifies such a situation. Like a Class-C amplifier, the Class-D amplifier involves pulsed operation, and requires an LC output circuit. However, that is where the resemblance ends. The Class-D operated device produces a pulse-width-modulated waveform which bears

Fig. 1-18. Input and output waveforms associated with Class-A, -B, and -C operation of power devices.

Fig. 1-19. The basic Class-D or "two-state" amplifier system.

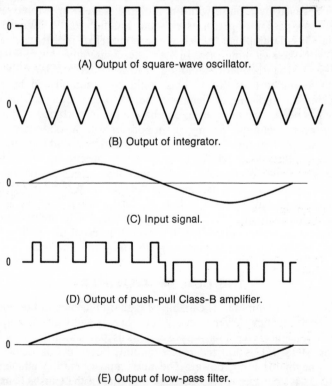

(A) Output of square-wave oscillator.

(B) Output of integrator.

(C) Input signal.

(D) Output of push-pull Class-B amplifier.

(E) Output of low-pass filter.

Fig. 1-20. Waveforms from Class-D amplifier system in Fig. 1-19.

the information extracted from an input signal. The LC circuit is a low-pass filter with sufficient damping or loading to *prevent* the type of ringing characteristic of the resonant tank of a Class-C amplifier. As a consequence, the output of the low-pass filter is a power-boosted version of the input signal. Viewed as a "black box," the Class-D amplifier functions as a linear amplifier. That is, one need not know about the strange things occurring between the input and the ultimate output of this amplifying system. In order to perform well, the low-pass output filter should do a good job of removing switching-frequency components. This is facilitated by making the switching rate much higher than the highest frequency to be amplified. Figs. 1-19 and 1-20 provide additional insights into the operation of Class-D amplifiers.

Sometimes the nomenclature Class D is applied to *other* switching-type circuits. The operation of an inverter, particularly a driven inverter, has been referred to as Class-D operation in technical literature. The switching device in the dc switching-type regulator has also been referred to as Class-D operation. Here, the resemblance to the amplifier just described is somewhat closer since the switching device is often keyed by a pulse-width-modulated waveform. Moreover, the output low-pass filter performs a similar function to the low-pass filter in the Class-D amplifier, which is the removal of frequencies associated with the switching rate. In fact, the commonly encountered 20-kHz switching regulator could be readily converted into an audio power amplifier (not hi-fi). This could be accomplished by superimposing the audio input signal on the reference voltage. This possibility has been cited in order to emphasize the importance of this relatively recent technique for processing power with a solid-state device. The desirable feature of switching-type operation is that the solid-state device is either on or off. When it is off, no output current flows, so there is no power dissipation. And when the device is on, only a minimal voltage drop exists between the "anode" and "cathode." Therefore, the power dissipation is *low*. If, furthermore, the rise and fall times of the switching transition is small compared to the width of the narrowest pulse being handled, the overall internal dissipation of the device will be low and its operating efficiency will be *high*.

THE CLASS-F AMPLIFIER

The Class-F amplifier is capable of higher efficiency in the boosting of radio-frequency power than are the Class-B and Class-C amplifiers traditionally used for this purpose. This is particularly true when a power MOSFET is utilized in the circuit, because this device consumes negligible input power. The maximum operating efficiency of the Class-B amplifier is 78.5 percent, and that of the Class-C amplifier is somewhat higher. However, ordinary Class-C amplifiers do not

Fig. 1-21. Basic Class-F radio-frequency power amplifier.

simultaneously develop maximum efficiency and maximum power output. In contrast, the ideal efficiency of the Class-F power amplifier is 100 percent. Inasmuch as Class-F operation is closely related to that of Class-B amplification, it will be more relevant to compare the Class-B and Class-F operational modes.

Fig. 1-21 is a representative circuit of a Class-F power amplifier. Because the MOSFET is an enhancement-type device, the biasing provision is automatically suitable for Class-B operation wherein output (drain) current flows for 180 degrees of the sine-wave input (gate) signal. Indeed, Class-B operation would occur, except for the effect of the quarter-wavelength transmission line inserted in the drain circuit. The effect of this unusual feature can be discerned from inspection of the waveforms in Fig. 1-22.

Fig. 1-22. Waveforms in the Class-F radio-frequency power amplifier.

In Fig. 1-22 the important waveforms are those of drain voltage and drain current. The dashed sine wave shown in the drain-voltage waveform depicts the situation which would prevail without the presence of the quarter-wavelength transmission line. This would then be Class-B operation.

Note that in Class-B operation there are times when we have both drain voltage and drain current. The simultaneous occurrence of both voltage and current results in power dissipation and is the reason for the theoretical limit of efficiency. In Class-F operation, the drain voltage is a square wave. As can be seen in Fig. 1-22, the drain voltage is zero when there is drain current. Also, the drain current is zero when drain voltage exists. This is a "pure" switching action; hence there is no internal dissipation and ideal operating efficiency is 100 percent. Of course, this ideal condition does not exist when the MOS-FET is operating in its linear, or power-dissipating mode.

It appears that the function of the quarter-wavelength line, in simplistic terms, is to convert the drain voltage from a sine wave to a square wave. This, indeed, is true. Inasmuch as the quarter-wavelength line is neither an active nor a nonlinear device, it is instructive to deduce the nature of its waveshaping property. In Class-B (and Class-C) operation, it is known that the resonant output circuit restores half-sine waves of a pulse train to a continuous sinusoidal waveform. How it accomplishes this can be explained in several ways. The explanation bearing greatest relevance to the topic under discussion is the selection and rejection of harmonic energy. Specifically, the parallel-resonant load circuit rejects all harmonics of the fundamental frequency. It does this by acting as a short circuit to these harmonics. At the same time, it *supports* the fundamental frequency, f_o. It does this by appearing as a high impedance to the fundamental frequency. (From the Fourier theorem of wave analysis, all nonsinusoidal waves are fabricated from sine-wave "building blocks". These sine waves are a fundamental frequency, f_o, and a number of harmonic frequencies, $2f_o$, $3f_o$, $4f_o$, $5f_o$, etc. The number, magnitude, and relative phases of the harmonics determine the shape of the wave. Only the sine wave, itself, is free of harmonics.)

Let us investigate what happens when the harmonic-discrimination properties of a quarter-wavelength transmission line are introduced into the circuit. For even harmonics, that is, $2f_o$, $4f_o$, $6f_o$, etc., the quarter-wavelength line behaves as a half-wavelength line. Such an element simply *reproduces* the load situation presented to the MOS-FET. Inasmuch as the parallel-resonant load supports f_o and short-circuits *all* harmonics, it follows that no change is thus far indicated. For odd harmonics, that is, f_o, $3f_o$, $5f_o$, $7f_o$, etc., the quarter-wavelength line displays the significant property that it *inverts* the impedance of the load, thereby allowing the odd harmonics of f_o to appear at the drain. Although f_o itself is, analytically speaking, an odd harmonic, the quarter-wavelength line does *not* reflect f_o at the load back to the drain. The differential treatment accorded to f_o stems from the fact that the load appears as a high impedance at f_o, while it appears as a short circuit to the *other* odd harmonics.

The *objective* is to impress a large number of odd harmonics of f_o at the drain, for such a combination, together with f_o, will synthesize the square-wave drain voltage needed for Class-F operation. So far, we have accounted for the accumulation of these odd harmonics, but it has been shown that the quarter-wavelength line will not return f_o itself from the load to the drain. Fortunately, this is not necessary because f_o already is present at the drain as represented by the dashed sine wave in Fig. 1-22. Accordingly, the odd harmonics returned by the quarter-wavelength line combine with the sine wave already present at the drain and produce the requisite square wave.

The shaded area between the two drain voltage waves in Fig. 1-22 is where dissipation takes place in Class-B operation. By causing the drain voltage to be a square wave, such dissipation is *eliminated*.

It might be supposed that the Class-F amplifier is only applicable to the uhf region, where a quarter-wavelength transmission line assumes practical linear dimensions. However, this interesting circuit can be implemented at lower frequencies within the amateur radio spectrum by using coiled coaxial cable. Coiling of coaxial cable will not appreciably affect its transmission line properties as long as the winding diameter is much greater than the diameter of the cable itself. Microdot miniature coaxial cable can be used to further reduce packaging size.

The Class-F power amplifier can be drain modulated. The parallel-resonant load circuit can be replaced by a pi network, or by a broadband transformer. With the Siliconix VMP-1 MOSFET, 11 or 12 watts of output should prove feasible. With the newer VN84GA 12.5-ampere, 80-volt MOSFET, several times this power level should be readily forthcoming. In any event, the main deterrent to ideal operating conditions is the ON resistance of the MOSFET. Unfortunately, this is not zero, as postulated in our concept of ideal switching action. Therefore, the realizable efficiency will tend to be in the area of 80 percent, at best. On the other hand, there is negligible loss associated with switching times. The Class-F amplifier, together with MOSFET power for radio-frequency amplification, is relatively new and is fertile ground for the experimentally inclined.

CLASS-G AMPLIFIER

Even the Class-B amplifier, or the Class-AB amplifier, cannot develop high efficiency when handling audio waveforms. Because of the high peak-to-average ratio of audio signals, and because low amplitudes prevail *most* of the time, the amplifier operates most of the time in its low-power region where efficiency is necessarily low. But, the amplifier must, nonetheless, be designed to accommodate the peaks. One naturally wonders whether something can be done about this situation.

If it were merely required to actuate an electric motor, it would be permissible to *clip* the peak excursions. Inasmuch as peaks are invested with relatively little energy, their absence would not be noticeable. Surprisingly, the same technique *can* be imparted to speech waves. Although the characteristics of the reproduction are altered, considerable clipping is allowable before *intelligibility* suffers. However, for high-fidelity reproduction of speech, and especially of *music,* peak-clipping is intolerable. Indeed, the ability to handle even occasional peaks explains the reason for 200-watt amplifiers which provide adequate volume in an average room while delivering less than a watt of output power.

The solution to this problem is found in the relatively new Class-G amplifier. With this amplifying system, a push-pull Class-B output stage has its dc supply voltage automatically switched in such a way that good operating efficiency prevails for both low- and high-amplitude input signals. More specifically, the output stage operates with low dc collector voltage when amplifying small signals, and with high collector voltage when amplifying large signals. Inasmuch as the optimum efficiency of Class-B amplifiers is attainable *only* when the collector voltage swing approaches the dc collector voltage level, the average efficiency of Class-B operation when processing audio signals falls considerably short of its potential. But when the Class-B stage is incorporated into the Class-G system, the average efficiency is much higher. The dividends reaped by this technique include various combinations of higher peak-power capability, reduced thermal problems, lowered manufacturing costs, and increased rms output power. The predominate feature depends to a large extent on the ratio of the high dc supply voltage to the low dc supply voltage. In practice, a reasonable blend of these features prevails when the low-voltage supplies provide about one half the dc voltage of their high-voltage counterparts.

In the simplified Class-G amplifier shown in Fig. 1-23, the Class-B output stage is the complementary pair of transistors, Q1 and Q3. Transistors Q2 and Q4 are *not* amplifying transistors, but serve as *switches* to transfer operation from supply voltage V1 to supply voltage V2 at the approximate time when the amplitude of the input signal exceeds the dc voltage level of the V1 power supplies. To understand how this occurs, consider only the operation of the upper portion of the circuit which includes transistors Q1 and Q2, diode D1, the positive V1 and V2 supplies, and the load. Initially, let the amplitude of

Fig. 1-23. Simplified circuit of the Class-G amplifier.

the input signal be *less* than V1. Under this operating condition, the load receives its current from supply V1 through diode D1 and transistor Q1. At the same time, transistor Q2 is reverse biased and is therefore inactive. Next, suppose that the input signal increases in amplitude to the extent that it exceeds the dc level of supply voltage V1

Fig. 1-24. Comparison in operating efficiency between Class-B and Class-G amplifiers.

(but it does not exceed the voltage level of supply V2). Under this operating condition, transistor Q2 becomes forward biased and turns on. Current from supply V2 now flows through the collector-emitter circuit of Q2, through transistor Q1, and then through the load back to supply V2. Thus, transistor Q2 has acted as an automatic switch to raise the dc collector voltage of Q1 when needed to accommodate higher power levels. As soon as transistor Q2 turns on, diode D1 becomes reverse biased, thereby *preventing* participation of supply V1. In its reverse-biased state diode D1 *also* prevents current from supply V2 from establishing a path back through the lower-voltage supply, V1. Fig. 1-24 illustrates the gain in efficiency over conventional Class-B operation.

Actually, when this technique is used in hi-fi audio applications, certain minor modifications are made. For, as one might suspect, the switching transition may not be as smooth as would be desired. In order to prevent, or greatly reduce, the approximately 0.6-volt "deadzone" due to the emitter-base contact potential of switching transistor Q2, a forward-conducting diode can be placed in the base lead of Class-B transistor Q1. Because such diodes also provide protection against excessive base-emitter reverse bias, they are often found in the base lead of *all* of the transistors. Because of charge-storage effects in the switching of transistor Q2 (and its counterpart, transistor Q4) a small *inductance* may be inserted in series with the load. However, these modifications do not materially change the basic operation as described. Also, the lower half of the circuit operates in the same manner as the upper half when the polarity of the input signal reverses. Note, however, that the polarities of Q3 and Q4 are reversed with respect to Q1 and Q2.

THE CLASS-H AMPLIFIER

The Class-H amplifier is an even more sophisticated solution to the tendency of conventional amplifiers to develop low efficiency with audio waveforms. Like the Class-G amplifier, the objective of the Class-H amplifier is to prevent application of more dc voltage to the output stage than is necessary to handle the instantaneous signal amplitude. For it is this *excess* of input dc power that causes unnecessary power dissipation—even the Class-B or -AB output stage develops its high efficiency *only* when the signal amplitude closely approaches the level of the dc voltage from the power supply. In many servoamplifier applications, continuous high efficiency can be provided by a Class-B output stage because the sine-wave signal is always at its maximum amplitude. As pointed out previously, audio waveforms are a far different from this situation because of their high ratio of peak-to-average amplitude. And because they operate for long periods at low am-

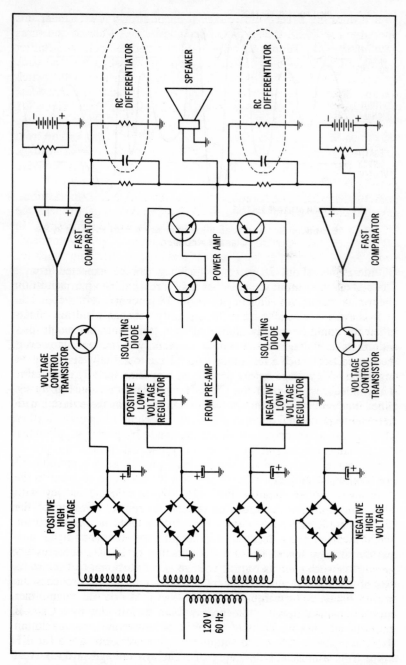

Fig. 1-25. Simplified circuit depicting basic concept of the Class-H amplifier.

51

Fig. 1-26. Anticipation and adaptation of power supply voltages in the Class-H amplifier.

plitudes, only about 20-percent efficiency can be expected from a Class-B or -AB output stage, rather than a reasonable approach to the theoretical maximum efficiency of about 78 percent.

It is evident, however, that the disparity between ideal and actual efficiency could be greatly alleviated by a technique that would provide high dc voltage to the output stage *only to the extent needed*. To be practical, such a technique would have to actually start increasing the supply voltage *before* the output amplifier actually required it. This, indeed, is just what the Class-H amplifier accomplishes. A significant advantage over the Class-G amplifier is that the output amplifier itself is not called upon to make any switching transitions. Rather, all adaptations to signal amplitude are made by power supplies that are actuated by the output signal amplitude. Therefore, the power-output stage is a relatively simple push-pull Class-AB circuit with conventional feedback.

In the simplified circuit of Fig. 1-25, voltage comparators with very high gains and slew rates turn on series-pass control transistors when output signals approach a set level. *Anticipation* of high-amplitude signal swings occurs because of differentiator networks in the comparator sensing leads. The transfer function of an RC differentiator network is such that its output voltage is proportional to the *rate of rise* of the input signal. It appears plausible that the sensing could be accomplished in a preamplifier, rather than in the output stage. Then, the propagation delay in the preamplifier, and in the driver stages, could be advantageously used to help achieve the anticipatory action. The anticipatory action and supply voltage adaptation in the Class-H amplifier are illustrated in Fig. 1-26. The low voltages are about 65 percent of the high-voltage levels.

2

Theories Pertinent to
Solid-State Power Devices

Working with *power* electronics is much like working with ordinary electronics except that there is a different *emphasis* on device characteristics. For example, such concepts as frequency response, current gain, and voltage polarity still retain their basic meanings. On the other hand, one must focus more attention on thermal impedance, charge storage, SOA (safe operating area), thermal runaway, and transients. And certain "simple" functions such as rectification or switching are found to exhibit complexities which, though of trivial importance in ordinary electronics, require careful attention in many power applications. For example, one might naively suppose that a certain low-power inverter circuit using SCRs can be simply scaled up to produce an inverter with, say, ten times the power-handling capability. But after increasing the ratings of all components including the SCRs by appropriate factors, it may be found that the high-power version of the circuit refuses to operate. Such peculiarities will be discussed in later paragraphs. To start with, let us investigate certain power aspects of the simplest of solid-state elements, the pn diode.

THE PHYSICS OF SEMICONDUCTOR BEHAVIOR

Our primary concern in this book is with *circuit applications of solid-state power devices*. This being the case, it is convenient to consider a device as a circuit component with input and output terminals. That is, we prefer not to become excessively involved with events transpiring *within* the device—the fact that such events occur gives rise to the parameters and characteristics of the device. Such a "black box" approach is quite effective in practice. However, much more insight into circuit action results from awareness of the basic nature of semiconductor behavior. For then we find ourselves in a *better* position to incorporate solid-state devices into appropriate circuits. Instead of merely assuming that a diode rectifies because it "passes current much more readily in one direction than the other," it will better serve our purposes to see *why* it exhibits this unilateral feature. It will then prove much easier to cope with certain behaviors of the solid-state devices that require special consideration in power electronics. Our involvement with semiconductor physics will be relevant, but minimal. Those desiring a more detailed study are referred to the specialized texts providing comprehensive treatment of this subject.

ELECTRICAL CONDUCTION IN METALS

The use of metallic wires to conduct electrical current generally does not produce a second thought when we connect the various components in a circuit. Yet, such conductors are also in essence components. Their conductive ability is as much a solid-state phenomenon as is transistor action in a semiconductor. Indeed, it may appear in retrospect that familiarity with the conduction of electrons in metals and the opposite behavior in insulators should have inspired earlier development of solid-state devices. This, of course, is an easy statement to make, but a tantalizing one too. For, in a qualitative way, it is true that diodes, transistors, and other semiconductor devices depend upon material that is *neither* a good conductor nor a good insulator, but rather somewhere in between—thus the prefix "semi." But, as we shall see, merely qualifying for classification as a poor conductor, or as a poor insulator, does not suffice for a material to be made into a diode or a transistor.

A metal such as copper is an elemental substance composed of copper atoms. The formation of the gross substance, copper, involves relatively tight packing of these atoms. Indeed, the atoms are thought to be so close together that their outer electronic orbits overlap as shown in Fig. 2-1. Such electrons are not uniquely bound to any one particular atom; rather these electrons exist in the material as a "fog" surrounding the atoms. Assuming that this description is substantially valid, it is next postulated that random motions are imparted to these electrons by the action of thermal energy. Interestingly, this erratic displacement of free electrons produces pulses of electric current which may be audibly or visibly detected with the aid of a high-gain amplifier. This so-called "noise" actually embraces a very wide frequency spectrum. However, no sustained flow of electricity takes place in such a copper conductor.

If we subject the free electrons to the pressure of an electric field by connecting a small battery across the ends of a copper wire, an

TYPICAL "FREE" ELECTRONS
AVAILABLE FOR CONDUCTION

$2 + 8 + 18 + 1 = 29$

TYPICAL COPPER ATOM

Fig. 2-1. Outer orbits of copper atoms share electrons.

orderly procession of electrons is *superimposed* upon the thermally induced random motion. Electrons are then supplied by the negative terminal of the battery and collected at the positive terminal. Within the wire, electrons are impelled to move from one atom to another. Other than thermally induced vibration, the bulk of the atom is *not* free to move. Also, it should not be thought that any particular electron injected into the negative end of the wire makes a speedy transition to the positive end. Rather, what we call a *flow* of electrons is more precisely a relaying process wherein the impetus received at the negative end of the wire is communicated with great speed to an electron at the positive end of the wire. The latter electron is ejected from the wire and collected at the positive terminal of the battery.

Thus, while it is true that electricity traverses such a conductor with a speed close to that of light, the motion of *individual* electrons from one end of the wire to the other is more in the nature of a slow drift. The basic mechanism involved may be demonstrated by arranging a row of marbles on a table. An impact conveyed to an end marble is quickly transmitted to the marble at the opposite end which is expelled from the row. The impacting "shooter" marble does not have to traverse the length of the row in order to eject the marble at the opposite end. Accompanying the ejection of the final marble was a "drift" of all marbles in the same direction toward which the final marble was ejected.

In our simplified model of conduction in metals, it is well to reflect that a *charge carrier* (the electron) is required. Although premature in the development of pn-junction theory, it is interesting to speculate the result of exerting control over the *availability* of the free carriers. The result would be some kind of a "solid-state" relay or amplifying device. Unfortunately, no practical way of doing this with an ordinary metallic conductor has evolved.

ELECTRICAL CONDUCTION IN LIQUIDS

Having accounted for electrical conduction in metals, we will next consider liquids and gases which will bring us somewhat closer to the conduction phenomenon in semiconductors. Let us now see why this is so. A liquid does not conduct in the manner of a metal because the atoms are not so densely compacted as in metals. Indeed, many liquids are excellent insulators for they are devoid of mobile charge carriers. The electron "fog" described for metals is not present in these liquids because orbital electrons in the atoms remain in captivity of the electrostatic forces of the atomic nuclei. Oils and alcohols generally belong to this category. Also, pure water tends to be an excellent insulator. Our usual conception of water identifies it with conductive properties. However, this is because we seldom encounter water with a high degree of purity. Because of the high dielectric constant of water, the electrostatic binding forces between the atoms of *other* molecules are weakened. The liberated atoms then become ions—that is, they bear electric charges. Such ions enable the solution to conduct electricity. Conduction in a solution of sodium chloride is shown in Fig. 2-2. If we are not interested in the conductive properties of the solution, or electrolyte, we can sum up the situation by classifying water as a good *solvent*. Fig. 2-3 illustrates the formation of crystalline sodium chloride. This atomic array is disassociated by dissolving the salt crystals in the water.

SODIUM CHLORIDE CRYSTALS
(TABLE SALT)

WATER

SODIUM CHLORIDE SOLUTION

Fig. 2-2. Although neither salt nor water alone provide charge carriers, their solution does.

THE CUBIC LATTICE

(+) = SODIUM ION
(−) = CHLORINE ION

BUILDUP OF GROSS SUBSTANCE
FROM REPETITIVE NETWORK
OF CUBIC LATTICES

Fig. 2-3. Formation of crystaline sodium chloride from repetitive arrays of ions.

A remarkable thing about many chemical reactions is that the participant atoms unite because of the tendency to have eight electrons in their outermost orbit. Somehow, the atoms then assume a more *stable* existence. During the course of the reaction electrons are given up, taken on, or shared until this stable condition is obtained. We then say that the reaction has gone to completion. An example touched upon was that of common table salt or sodium chloride. The sodium and chlorine atoms become *ions* in this reaction, but their final status is mutually beneficial because both find themselves with *eight* electrons in their outer orbit as shown in Fig. 2-4. We should not be surprised to find that the *inert*-gas atoms, neon, argon, krypton, and xenon *already* have eight electrons in their outer orbits (Fig. 2-5). The reluctance of these gases to participate in chemical reactions is due to the fact that such participation would tend to *deprive* them of their

TRANSFER OF ONE
ELECTRON

SODIUM ION   CHLORINE ION
BINDING FORCE DUE TO
OPPOSITE CHARGES OF
SODIUM AND CHLORINE IONS

Fig. 2-4. Electronic phenomena in the chemical reaction producing sodium chloride.

2, 8 (10+) NEON

2, 8, 8 (18+) ARGON

2, 8, 18, 8 (36+) KRYPTON

2, 8, 18, 18, 8 (54+) XENON

Fig. 2-5. The atoms of these four inert gases exemplify the stabilizing effect of eight electrons in their outer orbits.

stability. Let us now investigate the involvement of the magic number *eight* in the semiconductor material silicon.

An interesting aspect of electrical conduction in a liquid such as the sodium-chloride solution is that electrical conduction involves the movements of not just one charge carrier as in metals, but of *two*. The two carriers, the sodium ions and chlorine ions, bear *opposite* charge polarities and therefore move in *opposite* directions. Their contributions to the electric current are *additive*. This is an important idea to keep in mind, for it is relevant to the electrical phenomena in semiconductor material to be discussed shortly. It is interesting to contemplate that, whereas electrical current in the external wires connected to an electrolytic cell is derived from the movement of *electrons,* the internal current of the cell is *not* the consequence of electrons being transferred from one electrode to the other. Although the charges on the ions are due to a surplus or deficit of electrons, the *entire ions* move to the two electrodes where they are "collected." In this process, the ions again become atoms and, thus, we should expect to find a concentration of metallic sodium at the negative electrode and the evolution of chlorine gas at the positive electrode. (Actually, the sodium metal, being extremely reactive, quickly gets involved with other reactions in the solution. For example, some recombines with free chlorine and reforms the original salt or sodium chloride.) The concept of recombination should also be kept in mind, for a somewhat analagous occurrence is important in the operation of semiconductors.

ELECTRICAL CONDUCTION IN GASES

The physical characteristics of a gas are attributed to relatively large separations between the constituent atoms. Consider, for example, neon gas. Neon is classified as an *inert* element, implying a very weak affinity for other elements. This is so because the neon atom *already* has eight electrons in its outer orbit. So, it need not take on, give up, or share electrons with other atoms to improve its own stability. One might then suppose that neon gas is an excellent insulator. This is true; but paradoxically, the extensive use of neon in the electrical industry generally makes use of its *conductive* properties.

Neon gas, although a "natural" insulator for the aforementioned reasons, can be made to behave as a conductor by supplying it with a certain minimum energy. This added energy frees one or more of

Fig. 2-6. Conduction in neon gas.

the outer-orbit electrons from the attraction of the nucleus. When this occurs, the neon atom is *ionized* and can function as a charge carrier. The freed electron(s) is also available as a mobile charge carrier. Moreover, the ion and the freed electron may strike other atoms, thereby spreading the ionized state throughout the gas. Neon ions are ultimately collected at the negative electrode, whereas free electrons go to the positive electrode. However, both of their motions are *additive* with respect to the total electric current in the ionized gas. Electrical conduction in neon gas is depicted in Fig. 2-6.

The added energy needed to ionize neon gas can take several forms. For example, it is readily supplied in the form of an electric field. Raising the voltage across two electrodes situated in a neon-filled container ultimately attains a level sufficient to "breakdown," or *ionize,* the gas. In many practical devices, this occurs between 65 and 85 volts. Ionization is similarly produced by cosmic-ray particles, x-rays, and radio-frequency fields. Thermal energy can be supplied to either produce ionization or to facilitate its occurrence by other forms of energy. Electron beams, too, readily impart ionization energy. The common denominator in all cases is the *liberation* of one or more of the outer-orbit electrons in the neon atom. The study of electrical discharge in gases is complex and we have merely glossed over the basic principles. However, these highlights suffice to provide useful analogies with electrical phenomena in semiconductors.

THE ELECTRICAL NATURE OF INTRINSIC SILICON

Having discussed the several ways in which electrical conductivity can occur, or be brought about, we find ourselves in a more favorable position to deal with semiconductor material. Intrinsic, or *pure,* silicon at ordinary temperatures is a solid crystalline substance made up of silicon atoms. The individual silicon atom has four electrons in its outer orbit. This unique arrangement allows these atoms to unite with their *own kind* in the manner illustrated in Fig. 2-7. In such a formation, known as *covalent bonding,* each atom has attained the stable condition in which its outer orbit comprises *eight* rather than four electrons. As previously mentioned, eight outer electrons tend to confer stability to most atoms. Accordingly, pure crystalline silicon should be a good electrical *insulator,* since all electrons are retained in the force fields of the atomic nuclei. This prediction does not completely comply with observed behavior, however. Actually, there is a *low* conductivity in silicon at room temperature when photons force electrons in the outer orbit across the forbidden energy band into a unstable orbit. This conductivity increases exponentially with temperature and becomes significant at elevated temperatures. The implication of this is that the covalent bonds are not as strong as the binding forces

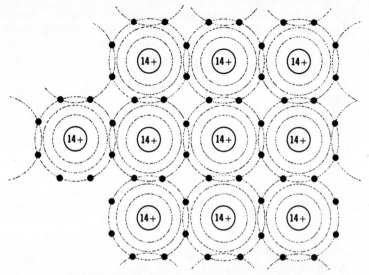

Fig. 2-7. Covalent sharing of electrons in silicon.

prevalent in the atomic structure of other insulating materials. Thus, the introduction of relatively little energy suffices to liberate occasional electrons which are then available to carry the electrical current. This is illustrated in Fig. 2-8.

At room temperature, silicon displays some of the aspects of metallic, liquid, and gaseous conduction. See Fig. 2-9. It resembles the behavior in metals in that the ionized atom itself cannot move through the substance. Intrinsic silicon is somewhat suggestive of a "weak" metal in which sparsely populated free carriers are available. The resemblance to liquids is the involvement of *two* charge carriers. One of these is the electron; the other is the vacancy, or "hole," left by a moving electron. The hole moves in the opposite direction to the electron and, despite the strangeness of the concept, should be thought of a free positive charge carrier. It contributes to current just as the more tangible positive ion does in liquids.

The analogy to gaseous conduction stems from the fact that energy must be added in order to free outer-orbit electrons from individual atoms. In any event, we do not yet have junction diodes or transistors resulting from the slight room-temperature conductivity of intrinsic silicon. Apparently, much further progress is needed in order to point the way for practical electronic control devices.

The Effect of Doping a Semiconductor Material

It is known that because of unique processing of intrinsic silicon, solid-state devices made from silicon can exhibit very high electrical

Fig. 2-8. At room temperature, a small force is sufficient to free electrons in silicon.

Fig. 2-9. Conduction in an intrinsic semiconductor material such as pure silicon.

conductivity, and that such conductivity is also subject to *control* by applied electrical signals. Let us now explore the means whereby such desirable action can be brought about.

The phosphorous atom is similar to the silicon atom, except that it contains *five,* rather than four, outer-orbit electrons. Suppose we add "just a pinch" of phosphorous atoms to crystalline silicon. Chemically and physically, any changes in the silicon would not, for practical purposes, be discernible. However, a singnificant *electrical* change would result. Fig. 2-10 shows the incorporation of the phosphorous or so-called *donor* atoms in the covalent bond structure of the silicon material. The situation is similar to that prevailing in "pure" silicon material in that all atoms are stable with *eight* electrons sharing the outer orbit. However, there is this *difference.* The fifth electron in the outer orbit of the phosphorous atom becomes a surplus electron. Therefore, such extra electrons are available in the adulterated or

Fig. 2-10. Donor atoms with five electrons in outer orbit contribute surplus electrons to silicon.

doped material as *conduction* electrons. Thus, the electrical conductivity of silicon semiconductor material is greatly increased because of the donor atoms and their extra orbital electrons. (Arsenic and tin, also with five outer electrons, merit consideration as donor atoms. However, the relative merits of each will not be discussed here.)

Silicon material doped with donor atoms is known as n-type semiconductor material because conductivity is enhanced via *negative* charge carriers, that is, the electrons.

63

P-Type Semiconductor Material

The logic of conductivity enhancement in n-type silicon material naturally leads to other speculations regarding processing of semiconductors. What, for example, would be the result of incorporating atoms such as aluminum with *three* outer-orbit electrons into the crystal lattice of silicon? If again, "just a pinch" of aluminum atoms are added, the physical and chemical characteristics of the bulk material (silicon) need not be appreciably affected. However, the electrical situation will then be as depicted in Fig. 2-11. The missing electrons, or *holes,* constitute deficiencies in the covalent bonding structure of the crystalline material. Because of this, there is a strong tendency for such holes to "capture" electrons. The electrons, in turn, are constantly being liberated from other atomic bonds by thermal energy. The net effect is that the conductivity of the semiconductor material is considerably enhanced by the p-type impurity, or *acceptor* atoms, and that the *main* current carrier is actually the hole. Since this is not easily visualized, it proves more convenient to simply think of the hole as a mobile *positive* charge carrier. The important points to keep in mind are as follows:

Fig. 2-11. Acceptor atoms with three electrons in outer orbit create "holes" in the covalent bonding structure of silicon.

1. In *both* n-type and p-type semiconductor material, electrical conductance is greatly enhanced as the consequence of *impurity* or *donor* atoms.
2. Although both negative and positive charge carriers participate in conducting the electrical current in both types of semiconductor, the *predominant* carrier in n-material is the *negative* electron. That is, electrons are *majority* carriers in *n*-material, and holes are minority carriers.
3. Conversely, in p-type material, *holes* are the majority carriers, and electrons are minority carriers.

THE PN JUNCTION

The characteristics of donor-doped and acceptor-doped silicon lead to one of the most important concepts in electronics. Specifically, it is the *pn junction,* which makes possible the rectifying diode, the transistor, thyristors, and various other solid-state devices, such as the zener diode, the varactor, and the junction FET.

Let us suppose that two separate regions, one doped with an n-type impurity and the other doped with a p-type impurity, are processed adjacently in silicon material. It is not easy to decide what phenomena should occur at the *junction* of these semiconductor regions. Indeed, one could make a good case for mutual annihilation on the one hand, or for *no* unusual action on the other hand. Actually, a unique formation of atoms does result. What happens is illustrated in four sequential steps in Fig. 2-12.

From this description of the formation of a pn junction one might visualize it as a reservoir of opposite charges paired off side by side. It would be natural to expect that such a condition could be only of a transitory nature, for it would appear that opposite charges must rush together and recombine. Even if this did not happen, we might suppose that the pn structure would behave as a charged capacitor, or as an electronic cell capable of providing current to an external load. By investigating the reasons that there is neither charge cancellation, chemical annihilation, nor an available current for an external load, we can gain valuable insights into mechanisms basic to the operation of solid-state devices.

Before dealing with the situations depicted in Fig. 2-12, it is well to be aware of the misleading representation of pn diodes so frequently found in technical literature. Thus, in Fig. 2-13A the pictorial symbol lacks an important feature of pn junctions, the depletion layer, as is shown in Fig. 2-13B. The depletion layer has the very low conduction properties of intrinsic silicon, even though the silicon material has been doped with impurity atoms. A simple way to account for this is to assume that equal numbers of oppositely charged impurity atoms

HOLES IN CRYSTAL LATTICE CREATE POSITIVE MAJORITY CARRIERS.

EXTRA ELECTRON IN CRYSTAL LATTICE CREATE NEGATIVE MAJORITY CARRIERS.

P REGION   N REGION

(A) Two adjacent impurity regions are formed in the silicon material.

(B) The majority charge carriers begin their migration toward one and another.

DEPLETION LAYER

ZONE IN WHICH MAJORITY CARRIERS RECOMBINE

IF THIS POSITIVE CHARGE MIGRATES TOWARDS JUNCTION, IT WILL BE TRAPPED BY FIXED NEGATIVE CHARGE.

IF THIS NEGATIVE CHARGE MIGRATES TOWARDS JUNCTION, IT WILL BE TRAPPED BY FIXED POSITIVE CHARGE.

(C) The plus signs indicate fixed positive ions and the minus signs indicate fixed negative ions.

(D) Thermally generated minority carriers maintain parallel arrays of opposite charges as well as a charge-free zone, or depletion layer, between them.

Fig. 2-12. The formation of a pn junction.

OHMIC CONTACTS

DEPLETION REGION

(A) A commonly encountered drawing.

(B) A more realistic drawing showing the depletion region.

Fig. 2-13. Representative illustrations of pn diodes.

exist in the central region of the pn structure, thereby producing a net cancellation of such charges.

A more realistic approach involves a dynamic, rather than static, concept of charge behavior. In Fig. 2-12A, imagine that the pn junction has just been formed. There is, indeed, a tendency for the electrons and holes provided by the impurity atoms to migrate across the junction and terminate their existence by recombination. However, there is a secondary effect in that the impurity atoms which provide these migratory electrons and holes are left in an ionized state, bearing charges *opposite* to the departed carriers. In Fig. 2-12B we see the majority charge carriers beginning their migration toward one another. Keeping in mind that the ionized impurity atoms are locked in the crystal lattice and are therefore *not* mobile, we see an interesting pattern forming in Fig. 2-12C. Finally, in Fig. 2-12D, the impurity ions form parallel rows or "plates" of opposite charges facing each other across the depletion layer. We are reminded of a charged capacitor, with the depletion layer behaving as the dielectric. (The analogy appears to be a good one, for physics texts explain that the stored charges in a physical capacitor actually reside on the inner surfaces of the metal plates.) The strange dilemma that no voltage or current from this "charged capacitor" is available from the terminals of a practical pn diode will be explained shortly. Fig. 2-14 combines the ideas developed in Figs. 2-12 and 2-13. It shows the charge distribution in a pn diode.

So far, we do not know whether conduction takes place as the result of externally applied voltages such as are shown in Fig. 2-15. Let us consider the situation in Fig. 2-15B with *forward* bias applied. That is, the p region is made positive relative to the n region. The majority carriers of both regions are *repulsed* from the respective electrodes and *driven toward* the depletion layer, making it very narrow. Due to the energy acquired from the electric field, many of these carriers cross over into oppositely doped regions and are collected at the electrodes there. Although some recombination occurs in the depletion region, many of the charge carriers, which now have directed motions, pass one another, continuing their journeys until arrival in the oppositely doped region. Such arrival manifests itself as current flow in the external circuit, much as plate current in a vacuum tube results from electrons arriving at the positively charged plate.

An interesting aspect of such conduction is that the p and n regions need not be uniformly doped with their respective impurity atoms. If, for example, the diode is made with relatively heavy impurity concentration in the p region as compared to the n region, conductivity will be primarily achieved via *holes* because of their availability. The converse situation prevails if the n region is made with relatively high impurity concentration. The conductivity, though still due to *majority*

DEPLETION REGION RESULTING
FROM RECOMBINATION OF
MOBILE CHARGES FROM IMPURITY
ATOMS IN P AND N REGIONS.

N REGION   P REGION

⊕ IMMOBILE POSITIVE CHARGES ON
IMPURITY ATOMS DUE TO LOSS
OF ELECTRONS

⊖ IMMOBILE NEGATIVE CHARGES
ON IMPURITY IONS DUE TO LOSS
OF HOLES.

·+ MOBILE POSITIVE CHARGES
(HOLES) CONTRIBUTED BY
IMPURITY ATOMS IN THE P
REGION.

– MOBILE NEGATIVE CHARGES
(ELECTRONS) CONTRIBUTED BY
IMPURITY ATOMS IN N REGION.

⊕⊖ CHARGE RECOMBINATION
(NEUTRALIZATION)

Fig. 2-14. The distribution of electric charges in a pn diode.

carriers, will then be primarily achieved via *electrons*. In all situations, forward bias causes *majority* carriers to be driven into a narrowed depletion region, and many of these carriers *penetrate* into the oppositely charged semiconductor region to be collected at the electrode. What happens if the applied bias voltage is reversed, as depicted in Figs. 2-15C and D? If for no other reason than that we have already

(A) With no applied voltage.

(B) With forward-bias voltage applied.

(C) With reverse-bias voltage applied.

(D) With a higher reverse-bias voltage applied than in (C).

Fig. 2-15. Behavior of the depletion region in a pn-junction diode.

labeled our device a "diode," we anticipate an entirely different phenomenon—one which *prevents* current from flowing into the external circuit.

When the p region of a pn junction is polarized negatively with respect to the n region, the mobile majority charges in the two regions are attracted to opposite charges appearing at their respective electrodes. Thus, electrons in the n region *recede* from the junction and move closer to the electrode contacting the n region. Similarly, holes in the p region *recede* from the junction and move closer to the electrode contacting the p region. The mutual action of the charges in the oppositely doped regions thereby widens the depletion layer. Significantly, current cannot flow by transport of majority carriers because the electric field tends to pull them *away* from the depletion layer. A very tiny current does flow due to thermally generated electron/hole pairs *within* the depletion layer. This is essentially the current that would flow in pure silicon. In summary, we have a device which provides thousands of times the conductivity in one direction of current flow than in the opposite direction. Thus, our device is a *rectifying diode.*

The Dilemma of the Internal Voltage

We have already discussed the paradox of the "charged plates" bounding the depletion layer in the pn junction. The question was raised: why is this voltage not detectable across the ohmic terminals of an ordinary pn-junction diode? Referring to Fig. 2-16, it is entirely acceptable to assume the existence of a voltage across the junction itself. However, by the time we get out to the *electrodes,* this voltage is no longer measurable. The situation is demonstrated by the series arrangement of cells shown in Fig. 2-16B. We see that the "internal voltage" is *not* available at the outer terminals of the cascaded cells. In the pn-junction diode of Fig. 2-16A, the relationship between the ions bounding the depletion layer and the majority charges in the rest of the semiconductor material is the same as that of the four cells in Fig. 2-16B.

Although this junction voltage is not available at the terminals of the pn-junction diode, it is very much in evidence since the forward-bias voltage applied to the diode must be on the order of 0.6 volt before the previously described forward conduction can take place.

The Reverse Current Is Not Always Trivial

We have qualified the pn diode as an exceedingly good rectifying device, mainly by virtue of the high ratio of forward conduction to reverse conduction. However, certain constraints must be complied with in order for the reverse current to be truly negligible. Referring to Fig. 2-17, we see that reverse current remains substantially constant

ZERO VOLTAGE AT TERMINALS

INTERNAL VOLTAGE

P

N

DEPLETION LAYER

+ = MAJORITY HOLES

− = MAJORITY ELECTRONS

⊕ = IMMOBLIE IMPURITY IONS CHARGED POSITIVELY BY LOSS OF ELECTRON

⊖ = IMMOBILE IMPURITY IONS CHARGED NEGATIVELY BY LOSS OF HOLE

INTERNAL VOLTAGE

ZERO VOLTAGE AT TERMINALS

(A) Static representation of the charge condition within a pn-junction diode.

(B) Equivalent battery circuit showing why internal "charged-capacitor" voltage does not appear across terminals of pn-junction diode.

Fig. 2-16. Why no voltage is detectable at the terminals of a pn-junction diode.

over a wide range of applied *reverse voltage.* However, reverse current is dependent upon *temperature.* The dependency is very strong, although this characteristic is not quantitatively revealed by Fig. 2-17.

AT HIGHER THAN ROOM TEMPERATURE

ROOM TEMPERATURE (21°C)

REVERSE CURRENT

REVERSE VOLTAGE

1000μA

100μA

10μA

1μA

0   5   10   15   20   25   30

Fig. 2-17. Reverse voltage behavior of a typical pn-junction diode.

If a pn diode is allowed to operate too hot, its rectifying action will be lost or it may be destroyed. Raising the temperature of a semiconductor material increases the number of thermally generated electron/hole pairs. Since these charge pairs are free to conduct electrical current, the material changes from a good insulator at room temperature to a good conductor at elevated temperatures. Indeed, if the temperature is high enough, the conduction resulting from the thermally generated charges overwhelms that resulting from the impurity atoms. Thus, the behavior of overheated silicon semiconductor material is essentially the same as for overheated intrinsic silicon.

THE PHENOMENA OF REVERSE VOLTAGE BREAKDOWN

Fig. 2-17 also indicates that a sufficiently high reverse voltage projects a pn-junction diode into an entirely different mode of conductivity, for the current abruptly undergoes a tremendous increase. For the example shown in Fig. 2-17, avalanche voltage breakdown occurs in the vicinity of 25 volts. We are reminded of the phenomenon of ionization in a gaseous device such as a neon bulb. The sudden breakdown is nondestructive as long as a series-limiting resistor is incorporated to restrict the maximum allowable current. The actual voltage at which breakdown occurs is governed somewhat by temperature, but much more so by the impurity concentration in the doped regions and by various geometrical and processing factors in the structure of the diode. The breakdown voltage is thereby controlled primarily by the manufacturer. This may range from several volts to several thousand volts. Actually, two different breakdown mechanisms are postulated, one known as avalanche breakdown and the other described as zener breakdown.

Avalanche Breakdown

According to this theory, the thermally generated charge carriers in the depletion layer obtain sufficient energy from the accelerating force of the applied electric field to dislodge other carriers from their covalent bonds. A cumulative production of carriers ensues wherein each available carrier impacts one or more lattice-bound charges with sufficient force to overcome the covalent bond. Such liberated charges not only become current carriers, but participate in the avalanche process by impacting with other lattice-bound charges. A tremendous number of carriers are created, and conduction becomes exceedingly high. This mechanism is analagous to the Townsend effect in electrified gas. The basic idea of avalanche breakdown is illustrated in Fig. 2-18. Note that the initiation of the process is dependent upon thermally generated carriers.

THERMALLY GENERATED ELECTRON-HOLE PAIR

R

Fig. 2-18. Avalanche breakdown in a pn junction.

Zener Breakdown

A second theory, that formulated by Zener, accounts for the current increase by assuming that the force of the applied electric field actually tears loose a large number of charges from their covalent bonds in the atomic structure of the crystal lattice. This process is depicted in Fig. 2-19. The difference between the two breakdown theories is somewhat subtle insofar as practical results are concerned. It has been determined, however, that the zener effect in silicon predominates at low voltages—below approximately 5.5 volts. (In germanium, it appears that the avalanche breakdown occurs first. That is why there were no germanium "zener" diodes on the market when germanium was the dominant semiconductor material.)

It has become customary to designate silicon breakdown diodes as zener diodes regardless of the breakdown mechanism. More significantly, every pn-junction diode, no matter what its intended function may be, is subject to breakdown when excessive reverse voltage is impressed across its terminals. And this also includes pn junctions within *other* devices, such as transistors, thyristors, and junction field-effect devices.

Reverse Breakdown Occurs With a Wide Depletion Region

It can be seen from Fig. 2-15 that the depletion layer *widens* with increasing reverse voltage. It has also been stated that the boundaries of the depletion layer are *fixed* ions. Therefore, the widening of the

72

Fig. 2-19. Zener breakdown in a pn junction.

depletion zone cannot be due to physical movement of these parallel
rows of boundary ions. What happens is that successively more dis-
tant rows of impurity atoms become boundary ions as the reverse
voltage is increased. A practical proof that this actually occurs will be
cited soon. But for the moment, it appears strange that reverse breakdown, whether it is an avalanche or a zener mechanism, takes place
in a relatively wide depletion region. If we liken the phenomenon to
a spark gap in a vacuum, our confusion deepens. But, it is to be noted
that the presence of gas atoms can impose voltage-breakdown characteristics in a gas tube which are not completely governed by mere
physical separation of the electrodes. And such is also the case with
reverse voltage breakdown in a junction diode. Despite the fact that
the depletion region becomes wider as the reverse voltage is raised, a
voltage is ultimately reached which suddenly is able to supply the
energy needed to generate a "cloud" of charge carriers. As previously
explained, this process abruptly liberates other carriers with the result
that the junction becomes highly conductive.

A MATHEMATICAL VIEWPOINT
OF PN-JUNCTION CURRENT

We have accounted for both forward and reverse conduction in
the pn-junction diode. One might easily suppose that entirely different
equations would be needed to describe the two currents since they

appear to invoke quite different conduction phenomena. In *forward conduction,* charge carriers are injected into a very narrow depletion region and manage to find their way across it to oppositely doped semiconductor regions where they are "collected," thereby producing current in the external circuit. Conversely, *reverse conduction* is a kind of "stray" effect which spoils what would otherwise be a high-quality insulating material. Specifically, thermally generated carriers within a relatively wide depletion region cause a tiny current. This current is nearly independent of the reverse voltage between about one-tenth of a volt and whatever voltage causes avalanche or zener breakdown in a particular diode. This contrasts with the *voltage-dependent* forward conduction current. Finally, reverse conduction is due to *minority* carriers only, whereas forward conduction involves both minority and majority carriers, but mainly *majority* carriers. With these differences in mind, it may come as somewhat of a surprise that a single, simple equation suffices to provide excellent quantitative data pertaining to the operation of a pn-junction diode in *both* operational modes—forward and reverse conduction.

This equation, moreover, can be deployed to automatically take into account the effect of temperature. And finally, it applies equally well to both silicon and germanium diodes. The equation is:

$$I = I_s(e^{40V} - 1)$$

where,

I is the current through the pn junction,
I_s isthe reverse current measured at temperature T,
V is the voltage across the diode.

Several observations are in order. We note that the value of temperature T need never be plugged into the equation. When V is reverse voltage, it is negative. A little experimentation reveals that for values of −V numerically greater than, say, −0.1 volt, the equation becomes for practical purposes $I = -I_s$ (recall that $e^{-x} = 1/e^x$). Finally, it can easily be ascertained that *positive* values of V above a few tenths of a volt will cause I to increase at an exceedingly high rate, at which time we have *forward* conduction.

It should be pointed out that this single equation is quite oblivious of the voltage where avalanche or zener breakdown occurs in any given pn junction. It is only able to tell us that current I will, ideally, be approximately I_s over a range of reverse voltage. However, the reverse breakdown voltage is ordinarily available from the manufacturer's specifications, or from direct measurement.

THE JUNCTION TRANSISTOR

As its name implies, the junction transistor derives its function from the behavior of the pn junction. Also implied by its name, the operation of the junction transistor involves the transit of a current from one region to another. Such an implication may appear to be trivial. However, it is well to bear in mind that if by some means a given current can be made dependent upon a *low* resistance, and then passed through a *high* resistance, *power amplification* can be achieved in the high resistance. From Ohm's law, a small change in current in the low resistance, R_1, will be accompanied by a power variation, P_1. This change in current will evoke a *larger* power variation, P_2, in the high resistance, R_2. This follows from $P = I^2R$. Unfortunately, if we merely connect two resistances in series in an attempt to prove this fact, our endeavor is defeated because the circuit current is primarily governed by the *high,* rather than the low, resistance. So, for the moment we are in possession of a tantalizing hypothesis, but one seemingly predicated more on fancy than fact. Obviously, there would be no point in indulging in fantasy—so let us investigate how actual realization of this power-gain mechanism is brought about in the transistor.

The transistor structure illustrated in Fig. 2-20A is seen to consist of two junction diodes fabricated in a single pellet of semiconductor material. It is shown connected to dc bias sources in such a manner

(A) Basic transistor circuit.

(B) Equivalent diode arrangement.

Fig. 2-20. Basic transistor structure and equivalent back-to-back diode arrangement.

that the left-hand diode is forward biased and the right-hand diode is reverse biased. This is inferred not only by observation of battery polarity, but also by the relative widths of the depletion layers. Note that the base region electrode is *common* to both "diodes." This arrangement is suggestive of the discrete diode circuit shown in Fig. 2-20B. A major *difference* between these two circuits is that in Fig. 2-20A, as we shall see, the current in the right-hand "diode" is very much determined by the current in the left-hand diode. In Fig. 2-20B there is *no* interdependency between the two diodes. Although the circuit in Fig. 2-20B depicts *some* aspects of the circuit in Fig. 2-20A, the resemblance proves superficial if we wish to simulate the power amplification possibilities of the transistor by the discrete diode circuit.

If, as has been inferred, the left-hand diode is capable of influencing the right-hand diode, it follows that the left-hand diode must behave as some sort of "charge gun" which injects current carriers into the right-hand diode. We might suspect such a mechanism by virtue of the reverse-bias voltage applied to the collector-base (right-hand) diode. If its behavior stemmed merely from its own characteristics, the collector-base diode would only support a tiny current in its external circuit and we would in essence have the back-to-back diode circuit of Fig. 2-20B.

THE SEMICONDUCTOR "ELECTRON GUN"

If we consider the emitter-base diode of Fig. 2-20A by itself, its function would be suggestive of a thermionic electron gun such as might be found in a cathode-ray tube. This is illustrated in Fig. 2-21. The qualitative analogy is good enough to provide further insights. It should be pointed out, however, that the electrons emerging from the accelerating grid or electrode in the thermionic gun have accelerations imparted by an electric field. This is not so in the semiconductor version, since the electrons which make their way across the base region are subjected to a virtually zero electric field. What impels

(A) Simplified representation of an emitter-base diode as a charge carrier.

(B) Analogous phenomena in a positive-grid vacuum tube or electron gun.

Fig. 2-21. The base-emitter diode can be considered analogous to an electron gun.

their transit more or less horizontally across the base region and why aren't they all "collected" at the bottom base electrode? The answers to these two questions are of key importance to the implementation of a transistor structure.

Note the zig-zag path indicated for the electron(s) in the p-type base region in Fig. 2-21A. This implies that electronic movement is due to *thermal agitation* and to the development of a so-called *concentration gradient*. A concentration gradient arises from the fact that like charges repel, causing a tendency for the electrons to continually seek regions of the base material in which electron population is less than in their present region. Such a mechanism helps *guide* the net movement of electrons across the thin base region. In other words, the electrons have a directed transverse movement *superimposed* upon their random motions. And, because the p-type base region is purposely made very thin, most electrons arrive at the far end region of the p-type material rather than at the base electrode. Such electrons are then susceptible to the attractive force of the collector electrode. (If it is not strictly proper to say that these electrons are "shot" into the collector region of the transistor structure, we can at least say that they are "introduced" there.)

You might suspect that this *largely haphazard migration* of electrons must limit the speed of response of the transistor. This, indeed, is true. However, processing technology has advanced the frequency range from tens of kilohertz in early power transistors to hundreds of megahertz in some modern designs. Some special microwave power transistors extend frequency response into the thousands of megahertz. As would be expected, one of the ways in which greater frequency response has been attained has been through techniques for forming *very thin* base regions.

By locating the base electrode on the side of the semiconductor "gun" as shown in Fig. 2-21A, the electrons are made to *enter* the base region, but are for the most part *not* collected by the base electrode. Another stratagem is involved in making the impurity doping in the n-type emitter region *much heavier* than in the p-type base region. This causes the predominant portion of charge movement to consist of electron injection from the emitter to the base region. Only a relatively small portion of charge movement consists of hole injection from the base to the emitter region. Electron collection by the base electrode tends to defeat the objective of the "gun" to make electrons available for the electric field in the collector region. Hole injection from the base to the emitter is also counter-productive for it only serves to increase external emitter base biasing current.

It is not to be inferred that such an electron gun as just described would work if placed in a vacuum. It certainly does function, however, when made an *integral part of the transistor structure*. We have considered it in "detached" form only for the sake of simplifying the discussion.

An interesting aspect of charge injection into the collector region of the transistor is that the charges coming out of the emitter region are *majority* carriers in *that* region. That is, they are the carriers resulting from impurity atoms in the emitter region. For the npn transistor structure under discussion, these charges are electrons and they become *minority* carriers in the base region because of the p-type semiconductor material. Now, when these electrons enter the n-type collector region they again constitute *minority* carriers. That is, they are the same type of charges that *already* are producing a small reverse current in the collector-base diode. It will be recalled that minority charge carriers are generated by thermal agitation and give rise to a relatively small current at room temperature. (Current due to minority carriers in any semiconductor region is simply the current that would exist without impurity atoms. Even though diode regions are doped with impurity atoms, reverse biasing produces only that current arising from thermal generation of *minority* carriers.)

From the foregoing discussion we have the interesting situation of a reverse-biased base-collector diode in which the tiny "natural" reverse current can be *greatly increased* by introducing a copious supply of current carriers from the emitter-base "electron gun." The prospects for amplification and power control now assume feasibility.

COMPARISON OF TRANSISTOR
AND VACUUM-TUBE AMPLIFIER

Our discussion of processes involved in junction transistor action is probably the minimum needed by those whose primary interest is solid-state circuits and applications. It should be obvious that those wishing to specialize in the physics and chemistry of semiconductors should study the comprehensive texts devoted to this phase of the technology. We shall now culminate our investigation of transistor action by directing attention to similarities existing in transistor and vacuum-tube amplifiers.

A comparison between transistor and vacuum-tube amplifiers is shown in Fig. 2-22. The npn transistor is depicted in a common-emitter circuit because this is the most popular circuit configuration. Another reason for dealing with the common-emitter circuit is because this is the way vacuum-tube amplifiers are most often used. It will be noted that the control grid of the tube is positively biased. This is done in order to reinforce the analogy and should prove acceptable inasmuch as it is permissible to operate tubes in this fashion under certain circumstances. For example, so-called "space-charge" tubes once used in automobile radios operated approximately this way. Also, the con-

(A) Transistor amplifier.   (B) Vacuum-tube amplifier.

Fig. 2-22. Comparison of the behavior of a transistor amplifier and a vacuum-tube amplifier.

trol grids of Class-B amplifier tubes are often driven into their positive regions. Actually, when a vacuum tube with the control grid positive with respect to the cathode is performing as an amplifier, the resemblance to transistor action is surprisingly close since there is grid-cathode current corresponding to base-emitter current in the transistor. Also, the input impedance of such a tube amplifier is relatively low, as is that of the transistor.

Table 2-1. Corresponding Regions Within Transistors and Tubes

Transistor Region	Function or Property	Vacuum-tube Region
Emitter region	Charges originate	Emitter or cathode
Depletion layer	Electric field region	Emitter-grid region
Base region	Control region	Grid
Depletion layer	Electric field region	Plate-grid region
Collector region	Charges collected here	Collector or plate

Table 2-1 will help correlate similar phenomena in the two devices shown in Fig. 2-22. Further insights may be gleaned from the analogies shown in Fig. 2-23. Here, not only are corresponding currents the same, but in both devices the current in the "input" diode is made up of negative charges flowing in one direction and positive charges flowing in the opposite direction. Significantly, in both cases the *net* cur-

79

(A) Transistor. (B) Vacuum tube.

Fig. 2-23. A more detailed example of analogous operation of a transistor and a vacuum tube.

rent measured in the control-electrode circuit is relatively small, being comprised of the *difference* between the currents in the collector (plate) and emitter (cathode) leads.

THE SILICON CONTROLLED RECTIFIER

The development of the bipolar transistor from the basic pn-junction diode appears natural enough in retrospect. But, surely, imagination and perhaps more than a trivial measure of luck were involved. The same has been true for *another* derivative of the pn junction—the *silicon controlled rectifier* (SCR). This time, one had the advantage that bipolar transistors could be used as "building blocks" for a very useful new device. Consider the simple transistor circuit of Fig. 2-24A. The outputs of the transistors, one an npn and the other a pnp, are connected to the inputs of one another. This arrangement is, no doubt, familiar enough since the first applications for the early junction transistors were in various oscillator and multivibrator circuits.

In any event, a momentary closure of switch Sw leads to a regeneratively reinforced latch-up condition with two current paths as depicted in Fig. 2-24B. The two transistors are turned on and remain heavily saturated because one current path begets the other. Indeed, this

(A) Simple circuit with npn and pnp transistors.

(B) Currents in circuit as consequence of momentary closure of switch Sw in (A).

(C) Pictorial representation of the transistors in (A).

(D) Semiconductor structure equivalent to the dual arrangement of (C).

(E) Schematic symbol.

Fig. 2-24. Building the silicon controlled rectifier.

cause-and-effect relationship is so persuasive that no signal applied to "G," the gate of the circuit, can turn the transistors off. Only by interrupting the power-supply current can the transistors be restored to their "off" conductive state. It is apparent that we have devised a good latching switch.

Fig. 2-24C is a pictorial representation of the transistors involved in the circuit of Fig. 2-24A. A *single* equivalent structure that can be made to substitute for this dual arrangement is shown in Fig. 2-24D. This consists of *four* semiconductor regions and *three* pn junctions. This is the basic structure of many thyristor devices including the SCR. Thus when fabricated on a single "chip," the thermal and electrical properties assume favorable values, and the economic and practical aspects also become attractive.

81

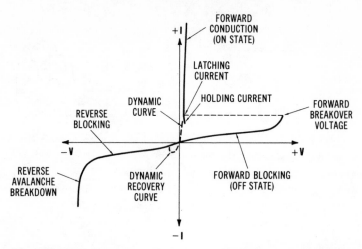

Fig. 2-25. Electrical characteristics of an SCR with zero gate voltage.

As a regenerative switch, the SCR and similar thyristors have revolutionized solid-state power electronics. When power is controlled by various switching techniques employing these devices, efficiency is very high because such switches are either on or off—never in intermediate or "linear" states where high power dissipation occurs. Of course, the circuit designer must exercise some ingenuity at times in order to control the duration of the on time and to commutate, or turn off, such devices in a desired fashion.

Fig. 2-25 illustrates the electrical characteristics of the SCR as a two-terminal device, that is, with no current delivered to the gate. Although the SCR is not commonly employed in this fashion, it provides insight to first consider it in this simple operating mode. Also, one would have to use the basic information in such an operating curve in order to specify an appropriate SCR for given circuit conditions. For example, if the circuit voltage equals or exceeds the forward breakdown voltage, the regenerative switching will take place without any gate signal. Also, the reverse voltage cannot equal or exceed the avalanche breakdown value. Otherwise, the SCR will no longer be an open switch for reverse voltage and will be forced to dissipate high, perhaps destructive, power. In addition, the load must permit sufficient holding current so that the SCR will remain conductive once triggered.

The effect of gate current is depicted in Fig. 2-26. The basic information presented in these curves is that *sufficient* gate current (at least I_{G4}) should be delivered to cause the SCR to abruptly switch from the blocking, or off, state to the on state over the entire anode-cathode voltage range. If the gate current is too small, the SCR will *not* be trig-

Fig. 2-26. The effect of gate current on the breakdown voltage of an SCR.

gered at low voltages. Sufficient gate current is also desirable to promote fast triggering, to provide a safety factor for the effects of temperature, and to ensure essentially similar circuit operation with different SCRs. The latter objective is generally met by using a trigger device such as a neon lamp, or preferably a semiconductor device such as a diac, a silicon unilateral switch, or a UJT.

An example of an SCR in a phase-control circuit is shown in Fig. 2-27. Here, the speed of a universal motor is controlled by varying the rms current through its armature and field winding. Although such control is efficient and effective for many applications, it should be borne in mind that top speed is essentially that which would prevail with the motor connected to the ac line through an ordinary silicon rectifying diode. If such a limitation is not allowable, the diode-switch circuit shown in dashed lines can be added. Ordinarily, the switch should be closed only when the SCR is operating at full conduction (approximately 180 degrees). However, the best way to achieve wide range continuous control is to use back-to-back SCRs, or a single triac.

Half-wave power control of motors, heaters, and lamps continues to merit consideration, in spite of the development of the full-wave thyristor, the triac. This is largely because SCRs remain superior when it comes to high power capability.

THE TRIAC

Although an SCR provides satisfactory control of ac power for many applications, there are very few instances where full-wave,

Fig. 2-27. Example of a simple SCR motor-control circuit.

AC SUPPLY	AC CURRENT	F1	CR1	R2	SCR1
120 V	1 A	3 AG, 1.5 A, Quick Act	RCA-1N3755	100 K, 1/2 W	RCA-2N3528
120 V	3 A	3 AB, 3 A	RCA-1N3755	100 K, 1/2 W	RCA-2N3228
120 V	7 A	3 AB, 7 A	RCA-1N3755	100 K, 1/2 W	RCA-2N3669
240 V	1 A	3 AG, 1.5 A, Quick Act	RCA-1N3756	150 K, 1/2 W	RCA-2N3529
240 V	3 A	3 AB, 3 A	RCA-1N3756	150 K, 1/2 W	RCA-2N3525
240 V	7 A	3 AB, 7 A	RCA-1N3756	150 K, 1/2 W	RCA-2N3670

Courtesy RCA Corp.

rather than half-wave, operation would not be desirable. In the full-wave mode, it is easier to obtain a wide range of power control. Even when the load is resistive, such as a heater or lamp, the power factor presented to the ac line is better (higher) with full-wave control. Moreover, if an isolating transformer is used, half-wave operation tends to produce core-saturation problems due to the dc component associated with this mode. Such an effect is not encountered with full-wave control. In certain inverter and power supply applications, filter problems are reduced in full-wave operation because the lowest harmonic present in the output is 2f, rather than f as with half-wave operation (f is the frequency of the ac voltage applied to the thyristors). Also, all *odd* harmonics except f, the *first harmonic,* are cancelled in the output of a full-wave control circuit.

A feasible method of utilizing SCRs in full-wave control systems is shown in Fig. 2-28A. Such a back-to-back connection is quite straightforward. One SCR is phase-controlled for one-half of the ac cycle and

(A) Basic concept.

(B) More detailed circuit showing special provision made for triggering the gates.

Fig. 2-28. The use of back-to-back SCRs to achieve full-wave control of load power.

the other SCR operates in symmetrical fashion for the alternate half-cycle. This results in full-wave control of load current. There is one drawback to this scheme insofar as concerns circuit simplicity in that the gates cannot be tied together. Rather, some isolation technique must be used so that the gate of each SCR can be triggered with respect to *its own* cathode. Additionally, the phase of the trigger signals must be displaced by 180 degrees even though their individual phases are varied in order to provide control of the load power. Finally, the two trigger-signals should be alike in amplitude, source impedance, rise time, etc. All of these requirements are readily met via the method shown in Fig. 2-28B. The gate control circuit is often a UJT relaxation oscillator.

Despite the feasibility of the basic scheme represented by Fig. 2-28B, it would be even better if a bidirectional thyristor could be devised, particularly one with a *single* control gate. Intuitively, one might suspect such a device would be possible because a given semiconductor geometry could be "shared" in time—in full-wave operation only one SCR in the circuits of Fig. 2-28 carries load current at any one time. And, inasmuch as the two SCRs are identical, why not utilize a *single* structure for such a control purpose? Indeed, such reasoning led to the development of the *triac,* an acronym for *tri*ode *ac* switch. It is also known as a *bidirectional triode thyristor.*

The symbol for the triac shown in Fig. 2-29A is intended to depict back-to-back SCR elements with a single gate electrode. In order to bring about the single-gate feature, the structure of the triac entails greater complexity than would result from simply combining two SCR structures. The terminals which handle the principle current through the device are labeled *main terminal 1* (MT1) and *main terminal 2* (MT2) rather than anode and cathode as in the SCR. The reason for this can be gleaned from inspection of the voltage-current characteristics of the device which are shown in Fig. 2-29B. Since there is no reverse voltage, symmetrical performance prevails for both po-

Courtesy RCA Corp.

(A) Symbol.

(B) Voltage-current characteristics of a triac with zero gate voltage.

Fig. 2-29. The triac is a full-wave thyristor which substitutes for a pair of SCRs.

larities of applied voltage. This feature, of course, is needed for full-wave operation. Accordingly, terminal MT1 is arbitrarily considered the "reference" terminal; the polarities of the gate and terminal MT2 are always designated with respect to MT1.

It is also true that the triac can be triggered by *either* a positive or a negative gate signal, whether terminal MT2 is positive or negative. Triacs are available with different gate sensitivities in order to accommodate various triggering modes. Many circuit applications favor triggering with a positive gate signal when terminal MT2 is positive and with a negative signal when terminal MT2 is negative.

The same need for gate trigger signals of *sufficient amplitude* as was described for the SCR also applies to the triac. Again, triggering will not occur at lower device voltages if sufficient gate current is not available. With sufficient gate current, the triggered triac will regeneratively revert from the blocking state to the conductive state. Once turned on

(A) Terminal MT2 positive and gate positive.

(B) Terminal MT2 positive and gate negative.

(C) Terminal MT2 negative and gate positive.

(D) Terminal MT2 negative and gate negative.

Fig. 2-30. The internal current paths in a triac.

in this manner, its characteristics will resemble those of a forward-biased silicon diode.

The triac is predominantly used in ac circuits and commutation occurs "naturally" as the current through terminals MT1-MT2 reverses direction. Triac operation is shown in Fig. 2-30 for the four possible combinations of polarity for terminal MT2 and the gate polarities. Figs. 2-30 A and D correspond to the mode of operation usually encountered in phase-control circuits. Note that the semiconductor "sandwich" is pnpn downward from terminal MT2 for one direction of the principle current, and is npnp downward from terminal MT2 when the principle current is in the opposite direction.

A typical circuit application of a triac is shown in Fig. 2-31. Besides providing wide-range full-wave control, this circuit features a dual-

	120VAC, 60Hz	240VAC, 60Hz
R1	2.2KΩ, 1/2W	3.3KΩ, 1/2W
R2	100KΩ, 1/2W	200KΩ, 1W
C1 C2	0.1μF, 200V	0.1μF, 400V
Y1	RCA 40485 } or 40431	RCA 40486 } or 40432
Y2	RCA 40583	RCA 40583

Courtesy RCA Corp.

Fig. 2-31. Typical triac lamp-control circuit.

time-constant phase-shift network to discourage hysteresis in the setting of R2, a diac triggering device, and an rfi filter.

THE JUNCTION FIELD-EFFECT TRANSISTOR

Some of the hardware applications of the pn junction described
have included the rectifying diode, the transistor, the varactor diode,
the zener diode, and the thyristor. There is at least one other important implementation of the pn junction. Until recently, the junction
field-effect transistor (JFET) was found only in small-signal and low-
power applications. However, these devices, along with insulated-gate
FETs (MOSFETs) are now serving as output devices in stereo amplifiers, in electric motor controls, and in other power systems requir-
ing control of tens or hundreds of watts. Insight into the principle of
the junction FET is readily attained by recalling the width modulation
of the depletion layers in varactor diodes and in transistors. If one
can control the geometry of the depletion regions by a voltage (as
represented by an electric field), it should be possible to thereby vary
resistance to a current. The situation should be somewhat analogous
to changing the cross-sectional area of a conducting wire. Unfortunately, the structure of diodes and transistors do not readily lend themselves to such behavior.

(A) Low reverse bias applied to gate.

(B) Moderate reverse bias applied to gate.

(C) High reverse bias applied to gate.

Fig. 2-32. Basic operating principle of the junction FET.

Fig. 2-32 illustrates the basic concept of a simple junction FET. The idea is to provide a length of semiconductor material with a central region where constriction of the effective cross-sectional area can take place. The means of producing such a constriction is by using a reverse-biased pn junction. The control voltage of the device is applied to one side of the pn junction through a terminal appropriately known as the *gate*. Note that the load current is controlled by the effective ohmic resistance of a path through the semiconductor material and does not pass through any pn junction, as in a bipolar transistor. (Accordingly, this device has been called a *unipolar transistor*.) Recent versions of FETs intended for power applications are fabricated to cause the controlled current to follow vertical, rather than horizontal, paths. However, the basic operating principle remains essentially the same.

THE INSULATED-GATE FIELD-EFFECT TRANSISTOR

Inasmuch as the gate junction of the JFET power transistor is actually deployed as a capacitor, it would be only natural to consider the feasibility of using an actual physical capacitor as a means of introducing an electric field into the semiconductor channel. Such a device is indeed practical as shown by the general configuration of the insulated-gate FET (IGFET or MOSFET) illustrated in Fig. 2-33. The gate capacitor is comprised of a film of silicon dioxide as the dielectric, and a metallized layer as one "plate"—the other plate may be said to be the p-type semiconductor material of the substrate. The electrical action of this device can be seen to be similar to that of the previously described JFET. Along with junction field-effect transistors, the emergence of insulated-gate types as power-controlling devices is relatively new. This may be somewhat surprising to those who have been conditioned for many years to regard all FETs as inherently limited to signal-level applications.

Both the JFET and MOSFET thus far discussed operate in the *depletion mode*. That is, they conduct heavily without bias voltage applied between the gate and source. The application of bias voltage then reduces the current through the drain-source circuit. The MOSFET (but not the JFET) can also be made so that operation occurs in the *enhancement* mode. In such operation, forward-bias voltage must be applied between the gate and source in order to produce current in the drain-source circuit. This, of course, is similar to the situation prevailing in ordinary bipolar transistors. A significant *difference* is that the input circuit of the bipolar transistor consumes current, whereas the gate of the MOSFET makes no such demand from the input source. Indeed, input resistance of the MOSFET can be about 10^{15} ohms, which is even higher than that of the JFET gate.

(A) Low negative voltage applied to gate electrode.

(B) Moderate negative voltage applied to gate electrode.

(C) High negative voltage applied to gate electrode.

Fig. 2-33. Basic operating principle of the depletion-mode MOSFET.

(A) Low positive voltage applied to gate electrode.

(B) Moderate positive voltage applied to gate electrode.

(C) High positive voltage applied to gate electrode.

Fig. 2-34. Basic operating principle of the enhancement-mode MOSFET.

The principle of the enhancement-mode MOSFET is illustrated in Fig. 2-34. The operating mechanism is somewhat more complex than in the depletion-mode device. The lower "plate" of the internal capacitor is again the p-type material of the substrate. The fabrication of the device is such, however, that this p-type region interrupts the n-type current-carrying channel. Accordingly, without gate voltage, drain-source current is negligible. The application of positive voltage between the gate and source *induces* an n-type region in the p-type material, thereby producing a controllable current-carrying n-type channel as shown in Figs. 2-34B and C.

It may appear that there is variance in the effect of the electric field in the depletion-mode and in the enhancement-mode devices. In one case the semiconductor material becomes "depleted"; in the other case an *induced* n-type region appears. The two effects are, however, manifestations of similar phenomena and can be manipulated by the manufacturer when impurities are "blended" with the intrinsic silicon.

P-CHANNEL FETs

Hitherto, we have only considered n-channel FETs. Field-effect transistors are, as with bipolar transistors, available with two types of polarity. The n-channel devices correspond to the npn transistor insofar as concerns bias polarities, and p-channel FETs correspond to pnp transistors. Fig. 2-35 shows p-channel versions of the junction and insulated-gate FETs previously discussed.

In the simple examples presented, it might appear that symmetry prevails with respect to source and drain connections and that they can be interchanged. This was true with some small-signal devices, but is not generally the case with the power FETs which have asymmetrical structures because the substrate is internally connected to the "source" and because of integrally contained protective diodes.

The V-groove, or "vertical" type, MOSFET used in power devices is shown in Fig. 2-36. This structure operates in the enhancement mode. The drain region encompasses a large area and permits effective heat removal. A protective zener diode for the gate is monolithically fabricated in this structure.

The different types of field-effect transistors are summarized and compared to bipolar transistors in the block diagrams of Fig. 2-37.

CONSIDERATION OF THERMODYNAMICS

When it comes to the design and implementation of solid-state power systems, it is often the "thermal circuit" that requires the greatest consideration. The matter of heat removal is so intimately related to performance, reliability, size, and cost, that any attempt at design

(A) Junction FET.

(B) Depletion-mode MOSFET.

(C) Enhancement-mode MOSFET.

Fig. 2-35. P-channel versions of junction and insulated-gate FETs.

or evaluation solely on the basis of electrical considerations is destined to be misleading, to say the least. At the same time, it would be quite difficult to provide specific guidance for the thermal management of each of the many practical applications to be presented later in this book. This is because of the great variation in individual construction and installation practices. Probably the most practical approach is to develop an awareness of the principles of heat flow as they apply to individual situations.

The reference to the *thermal circuit* is more than a mere play on words. Indeed, we make use of a so-called "Ohm's law" in the selec-

Courtesy Siliconix, Inc.

Fig. 2-36. Geometry used in V-grove power MOSFETs.

tion of power devices, in the methods of mounting them, and in the techniques of heat removal. The basic relationship, which appears simple enough, is given by the equation:

$$\theta = \frac{T}{P}$$

where,

θ represents the quanity known as "thermal resistance" which is expressed in degrees centigrade per watt (°C/W),

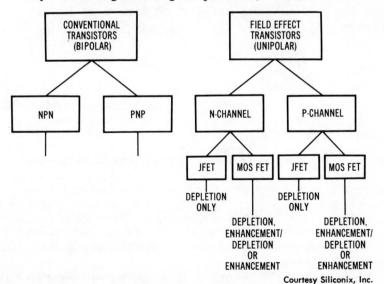

Courtesy Siliconix, Inc.

Fig. 2-37. Comparison between bipolar transistors and field-effect transistors.

T is the temperature rise in centigrade with respect to some reference temperature such as the ambient temperature,

P is either the allowable or resultant power dissipation in watts.

Of course, the algebraic permutations of this equation are equally useful. They are $P = T/\theta$ and $T = P\theta$.

Because of our natural liking for analogies, we welcome such an easy entry into the domain of thermodynamics. A closer examination of the simple relationship $\theta = T/P$ may invoke second thoughts, since we reflect that the electrical counterpart is $R = E^2/P$. From this, it must be inferred that temperature T is analogous to "voltage squared." Not only do we feel ill at ease with such a strange entity, but it is apparent that the "Ohm's law" of the thermal circuit is not really a one-to-one equivalent of the familiar electrical relationship. The explanation of this strange situation is not likely to be found in electronic literature, and requires considerable effort to dig it out of most physics textbooks.

It turns out that much use is made of thermal *conductance* in thermodynamics. As would be suspected, thermal conductance is $1/\theta$, or the reciprocal of thermal resistance. Thermal conductance is appropriately defined as the *flow rate of thermal energy per degree of temperature*. If we express this statement mathematically we should arrive at a clearer understanding of the way in which the so-called "Ohm's law" of the thermal circuit comes into being.

Energy in an electrical, thermal, or in any other system can be expressed as the product of power and time, or $P \times t$. Using the definition given above, we can now write our equation for thermal conductance:

$$\frac{1}{\theta} = \frac{\dfrac{P \times t}{T}}{t}$$

If we did not already have the definition for thermal conductance, an inspection of this equation would, indeed, tell us that thermal conductance is the *flow rate of thermal energy per degree of temperature*. Having thus derived this equation to comply with a basic concept in physics, we can simplify it by cancelling the t's, leaving:

$$\frac{1}{\theta} = \frac{P}{T}$$

Finally, if we invert this expression for thermal conductance, we arrive at our original "Ohm's law" relationship for thermal *resistance;* that is, $\theta = T/P$.

In the thermal circuit shown in Fig. 2-38, the resistances represent the thermal resistances associated with a power transistor and its heat

HEAT INPUT, P

JUNCTION TEMPERATURE, T_J

θ_{JC}

CASE TEMPERATURE, T_C

θ_{JA} θ_{CS}

θ_{CA}

HEAT SINK TEMPERATURE, T_S

θ_{SA}

AMBIENT TEMPERATURE, T_A

$T_J > T_C > T_S > T_A$

Fig. 2-38. Thermal circuit for a power transistor mounted on a heat sink.

sink. Note that power in the form of heat input is depicted as circulating in these resistances, and that certain temperatures exist at discrete points in the circuit. This makes the determination of important thermal quantities in a practical situation quite straightforward. Needed information is generally available from four sources: manufacturer's specifications sheets for solid-state devices, various tables and charts, electrical computations, and thermal "Ohm's law" computations. The usual procedure is best illustrated by a typical example.

In Fig. 2-38, θ_{JC} represents the thermal resistance from junction to case. This is found in the specification sheet and there is nothing we can do to change it. Here θ_{CS} represents the thermal resistance from case to heat sink, and θ_{SA} represents the thermal resistance from heat sink to ambient. The general objective is to make *these* two thermal resistances as low as skill, cost, and packaging considerations will permit. (When a heat sink is not used, θ_{CA}, the thermal resistance from case to ambient, assumes great importance. This thermal resistance is also often extracted from the manufacturer's specifications for the device.)

Suppose now that we are dealing with a voltage-regulated power supply using an RCA RCS29 power transistor as the series-pass element. Assume circuit conditions are such that maximum power dissipation occurs when this transistor is passing 2.25 amperes and has a voltage drop of 5 volts. The 11.25 watts of electrical power thereby dissipated is also 11.25 watts of "heat power." A destructive rise in junction temperature will occur unless a heat-removal technique is used. The mechanical and dimensional design of the power supply suggests that a heat sink in freely convected air (no fan or blower) might be suitable. How do we specify such a heat sink? The following steps typify the general procedure:

97

(1) T_J and θ_{JC} are recorded from the device specifications. For the RCS29 a junction temperature, T_J, up to 200°C is indicated, and the "built-in" junction-to-case thermal resistance, θ_{JC}, is 5.8°C/W.

(2) From Fig. 2-38 we can set up our thermal equation as follows:

$$\theta_{JC} = \frac{T_J - T_A}{P}$$

In this way, we shall solve for the total thermal resistance, θ_{JA}, in the circuit under operating conditions. (This is a necessary step in arriving at the thermal resistance of *one portion* of the circuit, specifically θ_{SA}, which is the heat-sink-to-ambient thermal resistance.)

In making the numerical substitutions it is usually best to use less than the maximum allowable junction temperature. In our example, we shall limit T_J to 175°C. Then, assuming a temperature of 25°C for T_A, we have:

$$\theta_{JA} = \frac{175 - 25}{11.25} = 13.3°C/W$$

(3) If we next obtain the two thermal resistances, θ_{JC} and θ_{CS}, we will be able to compute the maximum allowable thermal resistance from heat sink to ambient (θ_{SA}). It is then a simple matter of perusing through heat-sink catalogs to select a heat sink which is specified for θ_{SA}, or a lower thermal resistance. Table 2-2 shows commonly encountered values of θ_{CS} for

Table 2-2. Typical Case-to-Heat-Sink Thermal Resistances for Transistors With TO-3 and TO-66 Packages

Case	θCS				
	Metal-to-Metal		Using an Insulator		
	Dry	With Heat-sink Compound	With Heat-sink Compound	Type	
TO-3	0.2°C/W	0.1°C/W	0.4°C/W 0.35°C/W	3 mil MICA Anodized Aluminum	
TO-66	1.5°C/W	0.5°C/W	2.3°C/W	2 mil MICA	

Courtesy Motorola Semiconductor Products, Inc.

power transistors with the popular TO-66 and TO-3 metal cases. Our RCS29 transistor utilizes the TO-66 case. Assuming that a 2-mil mica insulating washer is to be used with a heat-sinking compound, it is seen that θ_{CS} is in the vicinity of 2.3°C/W.

(4) From $\theta_{SA} = \theta_{JA} - \theta_{JC} - \theta_{CS}$, we find that $\theta_{SA} = 13.3 - 5.8 - 2.3 = 5.2°C/W$. Referring to Table 2-3, we can choose from a number of heat sinks made by Wakefield, Thermalloy, IERC, and Staver. All of these manufacturers feature models with heat-sink-to-ambient thermal resistances in the 3.0 to 5.0°C/W range. For example, one could select the Thermalloy Type 6606.

Table 2-3. Heat-Sink-to-Ambient Thermal Resistance, θ_{SA}, for Some Commercially Available Heat Sinks

$\theta SA*(°C/W)$	TO-3 & TO-66
	Manufacturer/Series or Part Number
0.3-1.0	Thermalloy — 6441, 6443, 6450, 6470, 6560, 6590, 6660, 6690
1.0-3.0	Wakefield — 641 Thermalloy — 6123, 6135, 6169, 6306, 6401, 6403, 6421, 6423, 6427, 6442, 6463, 6500
3.0-5.0	Wakefield — 621, 623 Thermalloy — 6606, 6129, 6141, 6303 IERC — HP Staver — V3-3-2
5.0-7.0	Wakefield — 690 Thermalloy — 6002, 6003, 6004, 6005, 6052, 6053, 6054, 6176, 6301 IERC — LB Staver — V3-5-2
7.0-10.0	Wakefield — 672 Thermalloy — 6001, 6016, 6051, 6105, 6601 IERC — LA, uP Staver — V1-3, V1-5, V3-3, V3-5, V3-7
10.0-25.0	Thermalloy — 6013, 6014, 6015, 6103, 6104, 6105, 6117

Courtesy Motorola Semiconductor Products, Inc.

When fans or blowers are employed to provide forced air rather than natural convection, it is roughly equivalent to lowering the heat-sink-to-ambient thermal resistance, θ_{SA}. Although the improvement thereby realized is often appreciable, it is not always easy to calculate the effects of providing moving air. Much depends upon the nature of the motion (whether laminar or turbulent) and the air velocity in the immediate vicinity of the cooling fins on the heat sink. One can make use of the graphs and empirical data which abound in technical literature, but usually good fortune must prevail in order to adapt such information to a particular thermal situation. It is generally best to consult handbooks and manufacturers' literature for guidance, but be prepared for a good measure of experimental investigation.

In any situation, cooling occurs not only via conduction and convection but through *radiation* as well. It is usually rewarding to enhance this means of heat removal. This is particularly true because a prime consideration of radiation involves the nature of exposed surfaces. Otherwise knowledgeable people sometimes associate radiating efficiency with a bright and shiny surface. They probably think of electric heaters with mirrorlike reflectors. Such erroneous judgment stems from the overlooked fact that the heat is not developed *within* the metallic mass of the heater reflector as it is in a transistor and its heat sink. A so-called "black body" constitutes the ideal radiator, but a rough, dark surface often performs well. The radiating efficiency, or emissivity, of some practical surfaces is given in Table 2-4. Note

Table 2-4. Typical Emissivities of Common Surface

Surface	Emissivity, ϵ
Aluminum, Anodized	0.7 — 0.9
Alodine on Aluminum	0.15
Aluminum, Polished	0.05
Copper, Polished	0.07
Copper, Oxidized	0.70
Rolled Sheet Steel	0.66
Air Drying Enamel (any color)	0.85 — 0.91
Oil Paints (any color)	0.92 — 0.96
Varnish	0.89 — 0.93

Courtesy Motorola Semiconductor Products, Inc.

that surfaces with the best mirror qualities (highest optical reflectivities) tend to be the *poorest* heat radiators.

Heat removal by radiation requires that the substances expected to absorb the heat be at a lower temperature than the radiating body. Such absorbed heat would ultimately bring the heat absorbers to the same temperature as the radiator if it were not for the fact that the absorbers discharge their acquired heat to a larger environment external to the radiating and absorbing elements. In actual practice, the radiating body tends to lose heat until some equilibrium condition is attained in which both radiator and absorbers arrive at a constant temperature.

We find in physics textbooks that a hot radiating body propagates its heat energy into space at a rate proportional to the fourth power of its absolute temperature. Although it would appear that nature is on our side, there nonetheless remain pitfalls for the unwary. For example, one should not ignore the fact that the very same heat-sink surface selected for its high emissivity is also an excellent *absorber* of radiant heat energy. Thus, a transformer with a hotter surface temperature than that of a nearby heat sink will radiate heat *to* the heat sink. Such a situation will prevail even though the intervening air remains cooler than either the *radiator* (the hot transformer) or the *absorber* (the transistor heat sink) because of convection. Radiation can be employed to lower the temperature of "hot spots" *within* a package, but direct radiation into the *external* environment is usually a more effective use of this thermal technique.

3

Some Practical Aspects of Solid-State Devices

This chapter continues with the objectives to provide an appropriate preparation for the numerous applications presented in the final two chapters of this book. Practical considerations dictate that some of the ensuing discussion will be construed as review of the reader's arsenal of electronic facts. Some of it will be appraised to be nonrelevant by the "hardware-oriented" worker, while the more theory-inclined may experience frustration with treatments which do not pursue cause and effect relationships to their root sources. However, considering the basic insights to be found in these first three chapters, it is likely that various and useful benefits will accrue for those whose interest is stimulated by power electronics, notwithstanding their differing viewpoints.

DIODE RECTIFIERS

The basic specifications of current-carrying capacity and reverse-voltage rating are of paramount importance for rectifier diodes. However, there remains the extremely important parameter of *frequency response* or *switching speed* that is all too often overlooked. Although the diode rectifier is generally considered a passive element, it must be kept in mind that its essential function is that of a *switch*. This being the case, little concern was justified with regard to performance at 60 hertz, or even in the several-hundred-hertz region. With present-day power equipment utilizing frequencies in the kilohertz or tens-of-kilohertz range, we must take a second look at the "cut-and-dried" process of diode rectification.

(A) 60-hertz applied sine-wave voltage.

(B) 30,000-hertz applied sine-wave voltage.

Fig. 3-1. Low- and high-frequency behavior of "workhorse" junction diode.

A deterioration in rectification ability and efficiency sets in at higher operating frequencies. However, this is not the consequence of internal capacitance—at least not in the usual sense. That is, the current leakage through an internal capacitance of up to several tens of picofarads should not appreciably degrade rectification of a 30-kHz sine wave. Why, then, does the performance of some diodes deteriorate in the audio or the ultrasonic frequency range? Such impairment in rectifying performance, as illustrated in Fig. 3-1, manifests itself in internal dissipation and in higher ripple current in the external circuit.

Reverse-Recovery Characteristics

It was pointed out in the previous chapter that the current in a forward-biased junction diode was the net effect of *two* charge carriers, *electrons* and *holes* which are positive charge carriers. It

103

will be recalled that either of these carriers can be made to pre-dominate during the manufacturing process. The predominant, or majority, carrier is due to the impurity atoms in the semiconductor material and the minority carriers are generated by thermal agitation in the intrinsic atoms of the material. Although the behavior of these mobile charges and that of the fixed ionic charges produce the junction phenomena resulting in rectification, trouble is encountered when too much speed is expected from the minority carriers.

Specifically, if a junction diode is conducting during the forward-bias excursion of an ac voltage, numerous minority carriers participate in the current conduction. But, when the ac voltage reverses its polarity, many minority carriers *have not had ample time* to recombine with opposite charges and therefore remain "stored" in both the junction and in the bulk semiconductor material. Although these charge carriers are ultimately neutralized or collected, this takes *time*. Therefore, while such charge clearance is occurring, the diode *continues to conduct* despite the reversal of polarity of the impressed voltage. So, for a period of time, called the reverse-recovery time, the diode is no longer rectifying. Whether or not this lapse in rectification is serious depends upon how this interval relates to a half-cycle of the applied ac voltage. If the duration of the reverse-recovery time becomes excessive, the diode will heat up from the extra current load it is forced to carry. Moreover, the rectified dc output to the external circuit will be contaminated with increased ripple current and filter capacitors may be subject to temperature rise.

It can be appreciated that any diode which behaves in this fashion must ultimately experience trouble from low rectification efficiency and internally generated heat if the *frequency of the applied voltage is too high*.

Fig. 3-2 shows four reverse-current recovery characteristics commonly found in junction diodes. The magnitude, as well as the decay path, of the reverse current can be controlled by various processing techniques during manufacture. For such reverse current characteristics to be meaningful when two or more diodes are being compared, the measurements must be made under the *same conditions of forward current and of frequency*. Once this requirement is met, we can determine which characteristic is most desirable.

Other things being equal, the best diode would have no stored charge at all. Then rectification performance out to many tens or hundreds of kilohertz would be substantially as it is in junction diodes at 60 hertz. In practice, we must usually choose among such recovery characteristics as shown in Fig. 3-2. Note that the total reverse recovery time is depicted as t_{rr}. This is the time lapse from the initial zero crossing to a current value which is 25 percent of the maximum reverse current. (Although this is an arbitrary defi-

(A) Soft recovery with long total recovery time.

(B) Soft recovery with short total recovery time.

(C) Abrupt recovery with short total recovery time.

(D) Abrupt recovery with long total recovery time.

Courtesy Motorola Semiconductor Products, Inc.

Fig. 3-2. Various reverse-recovery characteristics in silicon junction diodes.

nition, it is a reasonable one.) Also, it is seen that t_{rr} is the sum of *two* intervals, t_a and t_b. The first interval, t_a, is due to charge storage in the depletion region of the junction. The second interval, t_b, comes about by virtue of charge storage in the bulk semiconductor material. The significance of this time resolution is that t_a denotes the time elapsing between zero crossing and the *peak* reverse current. For many practical applications, one needs to be concerned primarily with the total recovery time, t_{rr}.

It is also true that a relatively short total recovery time, t_{rr}, may not tell the whole story, since one must also know the value of the peak reverse current. In graphical terms, it is the area enclosed by the path of the recovery current that is meaningful. In technical terms, this area represents the total stored charge (or very nearly so). Total stored charge is important because it represents the energy available to produce circuit problems such as ringing, transients, degradation of rectification, and penetration of SOA curves in other devices. Therefore, a vendor should provide either numerical or graphical data relating to total recovered charge. Table 3-1 gives the stored charge in microcoulombs for some popular Motorola diodes. Note that the charge storage for the fast-recovery devices is *significantly less* than that for the ordinary diffused diodes. The fast recovery and low charge storage is brought about by a process of

105

Table 3-1. Stored Charge in Some Popular Motorola Silicon Diodes

Fast Recovery Devices		Diffused Devices	
Type	Typical Charge	Type	Typical Charge
1N4933 Series	0.08 μC*	1N4001 Series	2 μC
MR830 Series	0.12 μC	1N4719 Series	4 μC
1N3889 Series	0.16 μC	MR1120 Series	15 μC
1N3899 Series	0.19 μC	1N1183 Series	20 μC

These values are for a junction temperature of 25°C. The stored charge Q has a positive temperature coefficient of approximately 0.25% per degree centigrade.
* μC = microcoulomb

Courtesy Motorola Semiconductor Products, Inc.

gold doping during manufacture. This obviously represents a remarkable improvement in diode performance, particularly at higher frequencies. As may be expected, however, the reduction of charge storage imposes trade-offs in other diode parameters. Specifically, the gold-diffused silicon diode tends to have a *lower* reverse blocking voltage and a *higher* forward voltage drop than do ordinary diffused silicon diodes. The relative effects of gold doping on the reverse blocking voltage are illustrated in Fig. 3-3.

Although "common sense" dictates the use of the rapid recovery characteristic, some power electronic circuits may actually be adversely affected by *too abrupt* a return of the reverse current. This is because of ringing which may be shock-excited by such a rapid rate of change in current. The worst recovery characteristic for nearly all applications is one characterized by high energy storage together with abrupt returns. (High energy storage is usually indicated by high

(A) Diffused rectifier.

(B) Lightly gold-doped rectifier.

(C) Heavily gold-doped rectifier.

Fig. 3-3. Example of a parameter trade-off in gold doping of a silicon diode.

peak reverse current.) If, however, the energy storage is low, the abrupt return is generally favorable. Conversely, if peak reverse current is high, a softer return may be beneficial even though this infers greater energy storage. In many practical applications, it proves expedient to strive for low reverse current *and* a short t_{rr}. This, indeed, is what gold doping accomplishes.

Other Harmful Effects of Diode Charge Storage

The degradation of rectification is only one of the problems attending diode charge storage. Inasmuch as rectifier circuits are usually associated with other solid-state devices, it should come as no surprise that the functional reliablility of such devices may also be impaired. Moreover, simple rectification is only one function which diodes serve. Other functions include: free-wheeling diodes, SCR protection, charging diodes in LC ringing circuits, circuit-isolating diodes, commutation diodes, etc. The charge-storage phenomenon not only can cause faulty operation in these specific functions, but at the same time can adversely affect *other* devices and their functions.

Such an example is the simple switching circuit shown in Fig. 3-4. In this circuit, the transistor is doomed to failure unless a clamping diode is connected across the inductive load as shown. Note the col-

Fig. 3-4. A simple switching circuit using a clamping diode.

lector-current waveforms in Fig. 3-5. It is clearly seen that the clamping action of the conventional silicon diode is not nearly as good as that of the gold-doped, fast-recovery type. Even greater insight into the situation is conveyed by the load-line diagrams of Fig. 3-6. The entire upper portion of Fig. 3-6A represents needless dissipation in the switching transistor and could penetrate its SOA curve thereby causing failure. Of course, a lower switching frequency, a trapezoidal switching waveform, or a lower L/R ratio in the load tends to alleviate this problem and often enables satisfactory operation with the use of ordinary silicon diodes.

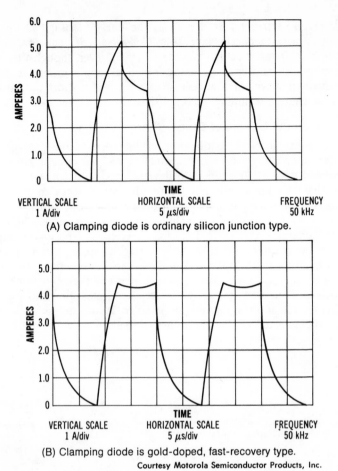

(A) Clamping diode is ordinary silicon junction type.

(B) Clamping diode is gold-doped, fast-recovery type.

Courtesy Motorola Semiconductor Products, Inc.

Fig. 3-5. Collector-current waveforms in switching circuit of Fig. 3-4.

Electrical Noise Considerations

Another diode characteristic which merits serious consideration in modern equipment is the rfi and emi potentiality of diodes. One must anticipate many high-frequency harmonics to be present in any source of energy that either turns on or off abruptly. Once these higher-frequency components are excited, it is only natural that some radiation will occur. There is certainly no dearth of "antennae" in the form of printed-circuit boards and interconnecting cables. Often, the high-frequency energy backs up into the *power line* from which it radiates only too effectively into other equipment.

It sometimes comes as a surprise that a circuit such as a bridge rectifier can be a vicious source of rfi and emi. This is the result of

(A) Clamping diode is ordinary silicon junction type.

(B) Clamping diode is gold-doped, fast-recovery type.

Courtesy Motorola Semiconductor Products, Inc.
Fig. 3-6. Load-line diagram for switching circuit of Fig. 3-4.

the recovery characteristic of the diodes and appears to again argue against a too abrupt recovery. However, the rewards of an abrupt recovery are a worthwhile objective in the audio and supersonic frequency ranges. One endeavors to suppress rfi and emi by suitable shielding and filtering techniques and by the use of appropriate components. For example, the use of toroidal-core components is very desirable in this regard.

Diodes With No Charge Storage

The previous discussion has concerned itself with the manipulation of the reverse recovery characteristic, or learning to live with it. Inasmuch as charge storage stems from the presence of minority carriers, the charge storage problem can be eliminated in a diode which accomplishes rectification with *majority* carriers only. Such a device is the *Schottky* diode. It consists of a metal in intimate contact with semiconductor material. With appropriate processing, this does not result in an ohmic contact; neither is a conventional pn junction formed. There is, however, a potential-barrier region at the interface that more or less simulates the behavior of the pn junction. One might view the Schottky barrier region as a half-junction. From the circuit viewpoint, two prominent features are provided by the Schottky diode. As previously mentioned, conduction takes place by majority carriers only and there is virtually no charge storage. Expressed another way, one might say that reverse recovery is instantaneous. (Keep in mind, however, that there is practically *no* reverse current.) The frequency capability of these devices is ultimately degraded by internal capacitance, but this effect does not set in until hundreds of kilohertz or higher.

The second compelling feature of the Schottky diode is its relatively low forward voltage drop. With the increasing demand for low-voltage, high-current dc power supplies, rectification efficiency has become very dependent upon this parameter. Fig. 3-7 shows a comparison between the forward voltage drops of otherwise similarly rated diodes—one being a conventional pn silicon type, and the other being a Schottky type.

We might suspect that the elimination of charge storage in the Schottky diode is possible only at the expense of other parameters. Fortunately, the forward voltage drop is not the trade-off, as it tends to be in the gold-doped type of fast-recovery diode. Indeed, we do find that the Schottky diode is not particularly good in two of its *reverse* ratings—it tends to have high leakage current and the maximum allowable reverse voltage is about 40 volts or so. This is not a problem when the device is exploited for its low forward voltage drop in low-voltage dc supplies. It does, however, limit its applications. When first available commercially these devices handled a few tens of amperes, were limited to reverse-voltage ratings of 20 volts, and "junction" temperature was rated at 100°C maximum. At this writing, Schottky diodes are available with current capabilities specified at 75 amperes (continuous), 40-volt reverse-voltage ratings, and at maximum junction temperatures up to 150°C. These rectifiers can be used with 20-kHz square waves, whereupon they will deliver clean, almost ideally shaped unidirectional pulses.

Courtesy Motorola Semiconductor Products, Inc.
Fig. 3-7. Comparison of forward current characteristics of a Schottky diode and a silicon pn junction diode.

The circuit segments illustrated in Fig. 3-8 are frequently encountered in solid-state power electronics. The diodes shown should be either gold-doped, fast-recovery types or Schottky diodes if the operating frequency or switching rate exceeds about ten kilohertz in most cases. The full-wave, center-tap rectifier of Fig. 3-8D is particularly suitable for Schottky diodes if the output voltage is to be five or ten volts. In this case, both the circuit and the diode can contribute maximally to high rectification efficiency.

THE BEHAVIOR OF SWITCHING TRANSISTORS

An inspection of a large number of power circuits will reveal that the use of transistors as switching devices is at least as important as is their classical employment as linear amplifiers, or as electronically controlled "rheostats." In theory, the process of switching can be 100-percent efficient because the *ideal* switching element exhibits no voltage drop across its terminals when in the on state, and involves no current leakage when in the off state. Since a real-life switch such as a transistor is nonideal in both of these respects, it cannot achieve lossless operation. However, there is at least another reason for losses within the switching transistor. Without knowing anything about the inner mechanism of the transistor, we would probably guess that it would not faithfully reproduce a square wave. The

(A) Free-wheeling diode in switching regulator.

(B) Isolating diode in flyback regulator.

(C) Isolating diode in boost regulator.

(D) Full-wave, center-tap rectifiers for high-frequency power supply with high-current, low-voltage output.

Fig. 3-8. Some suitable applications for Schottky diodes and gold-doped, fast-recovery diodes.

simple logic for this is that the vertical sides of a square wave require an infinite number of harmonics of the switching rate, and these harmonics must progress in diminishing amplitude to an infinite frequency. This being so, the most we could expect from a switching device is that it would support a large number of higher-frequency harmonics. But any departure from the ideal harmonic spectrum must necessarily degrade an impressed square wave into a trapezoidal-shaped waveform. The objection to the trapezoid is that the switching device is in its "linear" operating region when tracing out the sloped portions of the trapezoid and, therefore, incurs dissipative losses in the same manner as any rheostat.

Input and output current waveforms for a switching transistor are shown in Fig. 3-9. Note the "tail" on the input waveform. This is the reverse-current recovery characteristic of the emitter-base diode. During the entire period of reverse recovery, the collector (output) current behaves as if the emitter-base diode continued to be forward

Courtesy Motorola Semiconductor Products, Inc.

Fig. 3-9. Input and output current waveforms of a typical switching transistor.

biased. This indeed *is* the case, since it is true that a diode in reverse-current recovery *is internally* forward biased. (This is just another way of describing its charge storage.)

At low switching rates, the presentation of the waveforms in Fig. 3-9 would be nit-picking, for one would obtain an almost-perfect square-wave reproduction and the input and output currents would be time coincident for all practical purposes. However, when switching at the higher audio and the supersonic rates, the conditions depicted in Fig. 3-9 are no longer of trivial consequence. In addition, there is no guarantee that a 50/50 duty cycle will be reproduced, for it can be seen that the duration of the collector current waveform is at the mercy of the charge-storage behavior of the emitter-base region. Indeed, the region t_s is nonexistent if the transistor is not saturated. The length of t_s is proportional to the degree and time the transistor is driven into saturation. For most power control circuits, considerations of operating efficiency dictate a saturation mode with desirably low $V_{CE(sat)}$.

DRIVING POWER TRANSISTORS

A common requisite for all types of power transistors in both linear and switching circuits is proper *drive*. Such a statement might

(A) Delco DTS-409 npn tripple-diffused power transistor.

(B) Delco DTS-4067 npn tripple-diffused Darlington power transistor.

Fig. 3-10. Current gain variations in power transistors.

appear so obvious as to hardly merit discussion. However, inspection of such current-gain curves as shown in Fig. 3-10 clearly indicates why circuit designers generally must supplement rigorous design procedures with empirical tests and modifications. Note that the current-gain curves for the Darlington-type power transistor in Fig. 3-10B are plotted on a logarithmic scale and represent more extensive variation than does the plot for the ordinary npn transistor in Fig. 3-10A. The simple solution is to provide for *more* drive current than one might initially consider adequate at moderate temperatures and currents.

In switching circuits, the first requisite is usually to drive the power transistor well into its collector-saturation region. This is desirable in order to minimize $V_{CE(sat)}$, thereby reducing dissipation during on time. However, driving the base-emitter diode of the transistor *too* hard may prove self-defeating since the input losses then increase. Such emitter-base input losses contribute to the temperature rise of the transistor just as surely as does collector current. Moreover, excessive base current increases charge storage, which can manifest itself as a voltage *spike* in the collector circuit. Such spikes can be destructive because transistor operation is then projected outside the bounds of the SOA curve. Quite often, the spikes do not immediately destroy the transistor, but seem to be a definite factor in shortening the life of the transistor. At best, these voltage transients add to the electrical noise generated by the switching stage. The extent of this kind of malperformance is not ordinarily predictable due to its dependence upon stray circuit inductance.

In linear power control techniques such as amplifiers and dissipative regulators, the consequence of inadequate drive is generally quite obvious since the desired output power will not be available. The competent circuit designer must make adequate drive available under the worst conditions of temperature, line voltage, component aging, and load variation. In feedback systems, the phase characteristics of the driving circuitry assume importance, as well as voltage or current capability. Finally, in low-distortion amplifiers, it is always a better practice to design the driver for linear response than to use a high amount of loop feedback to compensate for the shortcomings of individual stages since the latter technique often leads to unstable or oscillatory operation.

THE SOA CURVES FOR TRANSISTORS

The term SOA is an acronym for *safe operating area*. The development of SOA curves for transistors has been one of the engineering milestones in the field of semiconductor devices. By the intelligent application of such graphical data, it becomes relatively

Fig. 3-11. Allowable collector current as a function of applied collector voltage.

easy to squeeze optimum performance from a power transistor with a sufficient safety factor for reliable operation. Prior to publication of SOA curves, one more or less "flew by the seat of his pants." There was always a mysterious aura surrounding transistor failure in power equipment, notwithstanding the tabulated specifications on allowable operating conditions. Such specifications are not readily correlated with the *dynamics* of actual transistor operation, even though one might argue that all needed information resides in the numerical limits indicated by the specifications. However, when a load-line diagram is superimposed upon the SOA curve, a quick visual inspection substitutes for many laborious calculations and interpolations. (Alternate terms for SOA are SRO or *safe region of operation,* and SOC or *safe operating curves.*)

Consider an ideal transistor for which the manufacturer claims that 50 watts of collector dissipation can be maintained at any combination of collector current and voltage. A simple graph of allowable collector current as a function of applied collector voltage is shown

116

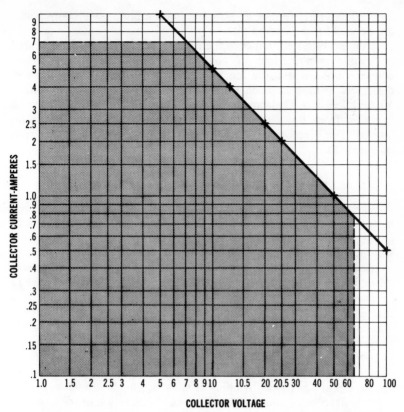

COLLECTOR CURRENT-AMPERES

COLLECTOR VOLTAGE

Fig. 3-12. Constant power hyperbola in Fig. 3-11 plotted on log-log graph paper.

in Fig. 3-11. Mathematically, such a curve is described by the equation $E \times I = 50$ (watts). This curve is a true *hyperbola*. An interesting and useful characteristic of the hyperbola is that it becomes a sloping straight line when the plot is made on log-log graph paper. Moreover, the slope is −1, or 45 degrees, no matter what numerical values of voltage and current generate the "curve," and no matter what the log-log scales are. This is shown in Fig. 3-12. The log-log plot has both graphical and numerical advantages over the curve plotted on ordinary coordinate graph paper. The log-log graph accommodates a greater voltage and current range and better graphical accuracy is obtained because the eye is quick to detect an erroneously plotted straight line. A curve could be quite deviant and yet not be obvious from a quick visual inspection. Also, if the curved plot is made to a different scale, its apparent shape changes, whereas the log-log plot remains a straight line with a 45-degree slope. One or

117

two points suffice to plot an accurate log-log plot, while many points are needed for the curved plot.

It should be noted that even the ideal transistor cannot be operated at infinite combinations of collector voltage and current. When collector voltage is made too high, avalanche breakdown occurs in the collector-base junction because of the excessive reverse voltage applied across it. Collector current is limited by the minority charge injection from the emitter-base section, which itself is dissipation limited. The graph in Fig. 3-11 clearly shows these allowable limits of operation.

TRANSISTOR BREAKDOWN CHARACTERISTICS

When power transistors were first employed for switches, inverters, motor controllers, and the like, there were many mysterious failures in the field. This was support for the contentions of the die-hards who maintained that semiconductor devices were "naturally" destined to prove unreliable. The failure mode was dubbed *secondary breakdown*, for it was found to resemble and often (but not necessarily) follow avalanche or *first breakdown*. A significant difference was that, in most circuit applications, the so-called secondary breakdown was much more likely to destroy the transistor. For awhile, little progress was made in achieving understanding of the destructive phenomenon. The problem could not qualify for solution until a straightforward and reliable rating technique was found that would enable circuit designers to apply it with justifiable confidence. This turned out to be an additional "off-limits" boundary line added to the graph of Fig. 3-12. The resultant graph then became popularly known as the SOA (safe operating area) curve. Fig. 3-13 depicts an actual example of this stratagem. It is seen that there are now four limits imposed on voltage, current, or certain *combinations of the two*. The added restriction of the SOA curve protects against occurrence of the once-mysterious secondary breakdown. In essence, SOA curves inform us that a transistor cannot operate at maximum current, maximum voltage, and maximum power at the same time.

The Two Types of Secondary Breakdown

With regard to secondary breakdown, it turns out that there are *two* distinguishable types. This is not splitting hairs, for one type of secondary breakdown differs from the other in the very important aspect that less energy suffices to trigger it. The two types of breakdowns are known as *forward-biased* secondary breakdown and *reverse-biased* secondary breakdown. The biasing refers to the emitter-base section of the transistor. Forward-biased secondary breakdown can occur while the transistor is operating in its "active" region; that is, collector current flows because of the emitter-base bias. Circuits which come to mind as examples are Class-A amplifiers and series-pass regulators.

In contrast, reverse-biased secondary breakdown can occur when the forward emitter-base bias has been removed (or reversed) and the collector current has thereby been switched off. Unfortunately, inductance in the collector circuit attempts to maintain the flow of

Fig. 3-13. SOA graphs for two similar RCA pnp transistors.

collector current which leads to the possibility of reverse-biased secondary breakdown. This tends to occur with lower energy levels than does forward-biased secondary breakdown. Contrary to the numerical implications, either breakdown can occur without the transistor having to first undergo avalanche breakdown. Examples of circuits vulnerable to reverse-biased secondary breakdown are inverters and switching regulators.

How Does Secondary Breakdown Differ
From Avalanche Breakdown?

Both types of secondary breakdown are thermally related in that they can be thought of as extremely fast thermal-runaway situations. Secondary breakdowns are caused by severe current concentrations occurring in small semiconductor areas which constitute *hot spots*. That is why secondary breakdown can happen even though the *average* transistor junction temperature is well within specified limits. Such hot spots inject more charge carriers into the collector region, thereby increasing collector current. This in turn "adds fuel to the fire," causing even greater temperature rise to be developed in the hot spots. The process is *regenerative* and initially involves such a small mass that the transistor can be destroyed before the average junction or case temperature undergoes appreciable rise.

Thus, secondary breakdown is a *thermal phenomenon,* whereas avalanche breakdown from excessive collector voltage is initiated by a high electric-field gradient. This has important practical ramifications even though the end result of both breakdown phenomena is high collector current. The final limiting voltage following the onset of secondary breakdown is generally in the 5- to 30-volt range. Avalanche breakdown voltage tends to be appreciably higher than this. Secondary breakdown occurs at certain *combinations* of *voltage, current* and *time.* The fact that *time* is involved shows that secondary breakdown is basically *energy-dependent.* It might appear that with *three* governing parameters, rather than just voltage and current, our ability to guard against secondary breakdown would be destined to remain a black art. That this is not so is clearly revealed by the set of SOA curves shown in Fig. 3-14. (This is actually more typical of published data than is the basic graph of Fig. 3-13.)

We see that penetration of the dc operation boundaries is permissible for *short* on times. The shorter the on time, the more the applicable SOA boundaries extend into *higher* voltage and current areas. This reinforces the previously made statement that secondary breakdown is *energy* dependent. Indeed, it is true that for *very short* on times, the secondary breakdown limit disappears from the SOA curves.

Interestingly, secondary breakdown can be made to take place without necessarily destroying the transistor. Transistor test sets used by manufacturers interrupt the power supplied to the transistor so quickly at the onset of secondary breakdown that no degradation to the transistor occurs. Designers often build such test equipment to better correlate their results with the true transistor characteristics. However, this is not generally necessary. The quantitative accuracy of published SOA curves is quite good and may be considered reliable

Fig. 3-14. SOA graph for General Electric D27C and D27D power transistors.

for most engineering purposes. Admittedly, some skill must be cultivated in order to effectively translate graphical data to practical hardware. Deviations from true standardization are all too common among manufacturers, which does not make the task easier. All things considered, however, SOA curves provide reasonable predictability for the operation of power transistors. Bear in mind that most SOA graphs pertain to forward-bias operation. If an inductive circuit is switched off, one should press the vendor for SOA data in the reverse-bias operating mode since the allowable energy absorption may be appreciably less than for forward-bias operation.

Temperature Derating

The SOA curves shown in Fig. 3-14 are valid for transistor case temperatures up to 70°C. In practice, this usually means that if a suitable heat sink is used, no temperature derating of the SOA curves is necessary. Often, however, the SOA curves are given for a case temperature of 25°C. It is not likely that one can use a large enough heat sink to maintain the transistor case temperature at 25°C, which is only slightly above "room temperature." In order to adapt to practical operating situations, such 25°C SOA graphs are accompanied by temperature derating curves as illustrated in Fig. 3-15. Before the SOA graphs can be used, the dissipation-limited boundary lines must be lowered to comply with the actual case temperature. This is also true for the $I_{s/b}$ line as well. The case temperature may be derived from calculations of the thermal circuit, or may be estimated. Experience plus willingness to experiment somewhat with thermal hardware is usually necessary whether or not thermal calculations are made, for there are often variables associated with the packaging situation that are not easy to measure or evaluate. Note that the choice of a higher-than-actual case temperature is a conservative approach which allows some safety factor.

THERMAL CYCLING OF POWER TRANSISTORS

Although there are no filaments to cease functioning, vacuums to become gassy, or contacts to burn and pit, one cannot enjoy the

(A) SOA graphs.

Fig. 3-15. 25°C SOA graphs and

luxury of ignoring the life span of *power-level* solid-state devices. Along with lamps, batteries, and electrolytic capacitors, it is often necessary to ascertain that reasonable service will be obtained, even *after* proper attention to SOA ratings. (Defiance of SOA ratings is often the cause of immediate catastrophic failure.) Surely, we had hoped that compliance with SOA ratings and temperature bounds would enable the transistor to last as long as circuit boards and heat sinks. What, indeed, "wears out" in an operating transistor not subjected to transients or other electrical abuse?

It turns out that the life-depleting agent in power transistors is *thermal fatigue*. Somewhere or sometime we have probably encountered this term but, because it has often had vague or nebulous connotations, the usual reaction to it has been that it has little place in the listing of predictable phenomena. For example, one might ascribe the unexpetced blowout of a fairly new automobile tire to thermal fatigue. Similarly, a burned-out light bulb can be conveniently given such an autopsy. Because the life spans of tires and lamp filaments are quite variable, our accounts of their demise are not always convincing. Surprisingly, however, the proper consideration of thermal-fatigue data as applied to *power transistors* can project a reasonably accurate limit to anticipated life span.

Thermal fatigue occurs in power transistors because of differential expansions and contractions of the silicon pellet, the case, solder or

(B) Derating chart.

Courtesy RCA Corp.

accompanying derating chart.

weldments, and the bonds used in the construction and packaging of the device. The manufacturer can do his utmost to minimize these thermomechanical displacements, but he cannot eliminate them entirely. It has been found that the disruption of transistor operation by thermal fatigue is a predictable function of the *number* of thermal cycles that the transistor is subjected to and the temperature excursion of these thermal cycles. Fig. 3-16 shows the thermal-cycle rating chart for the RCA RCS258 npn power transistor. Obviously, the circuit designer should strive for effective heat removal and, if feasible, should minimize the number of anticipated temperature cycles.

SELECTION OF TRANSISTOR TYPES

It is quite easy to become confused in the selection and evaluation of power transistors. Not only is there a bewildering array of types, classifications, and fabrications, but the performance characteristics allegedly innate to a particular kind of transistor may vary from manufacturer to manufacturer. A classical example is the long popular workhorse power transistor, the 2N3055. This device can lay claim to no thoroughbred certification, since its construction, geometry, and chip size has literally varied all over the map. Also, it is *not* common for a manufacturer's technical literature to play down the current and voltage capabilities of their products, even when the apparently superb performance pertains to unrealistic situations in practice.

Fortunately, basic guidelines for the selection of transistors do exist. The very fact that considerable overlap prevails is an asset rather than a liability to the designer's needs. The generalized listing of power transistor applications presented in Table 3-2 can, at the least, help exclude types likely to be unsuitable. In this table, a *low-power* transistor is one capable of dissipating from 1 to 10 watts, a medium-power transistor is specified for 10 to 100 watts of dissipation, and 100 watts or more of dissipation qualifies a device as a high-power transistor. If these and the classifications to follow appear arbitrary, they are nonetheless valuable from a relative viewpoint.

Current capability, like power dissipation, is often a ploy of the specification sheets since no rating is absurd if the necessary heat removal can be practically or economically implemented. With this in mind, it becomes useful to consider power transistors with less than 3 amperes of continuous load capability as low-current devices. Medium-current transistors are those capable of continuously supplying load currents from 3 to 10 amperes. And high-current transistors are those with continuous current capabilities exceeding 10 amperes.

Table 3-2. Generalized Applications of Power Transistors

Type	Features	Limitations	Applications
Germanium PNP	Low cost. High current capability. Low collector-emitter saturation voltage	Limited frequency capability. Effective heat removal needed. Collector voltage rating is generally below 100 volts. No npn.	Series-pass elements in linear regulators. Inverters, converters, and switching regulators operating at low and medium frequencies. Audio amplifiers, motor control.
Silicon Single Diffusion NPN	Electrically rugged. Superior thermal characteristics often make this device a better choice than germanium.	Limited frequency range and moderate voltage ratings. Not commonly available in pnp.	Linear regulators and medium-frequency switching applications. Audio amplifiers, motor control, solenoid drivers.
Silicon Epitaxial Base NPN/PNP	Rugged wtih good frequency range and voltage ratings. Useful for many power-control applications.	Possible marginal performance in high-frequency switching circuits.	General-purpose power device. Audio amplifiers, auto ignition, solenoid drivers.
Silicon Double Diffusion NPN/PNP	Best bipolar device for frequency range and high switching speed. Good current rating.	Ruggedness suffers at higher frequencies. SOA tends to be low.	High-speed switching applications, often used with protective techniques. Rf amplifiers.
Silicon Triple Diffusion NPN	Second-best bipolar device for frequency range and switching speed. Best voltage rating.	Current rating generally lower than other silicon transistors. Cost tends to exceed other "workhorse" types.	High-voltage regulators, inverters, and converters. High-voltage and high-speed switching applications. Rf amplifiers.
Silicon Darlington NPN/PNP	High current gain and high input impedance. Favorable cost and production characteristics.	Tradeoffs in frequency, voltage, and current capabilities must be carefully evaluated.	Series-pass elements, drivers, moderate-speed switchers, audio amplifiers, auto ignition, general-purpose power control.
Power MOSFET N-Channel	No thermal runaway or secondary breakdown. Very high switching speed and input impedance. Internally protected gate.	Current and voltage ratings lower than bipolar devices. "On" resistance may cause excessive voltage loss.	Low-power linear and switching-type power supplies. Audio amplifiers, rf amplifiers, logic-driven power control.
Synthesized Transistor	Overload protected. Base drive as high as 40 volts without damage. 0.5-μs switching time. 3-μA base current. Interfaces CMOS or TTL.	Limited current and voltage ratings. (Approx. 2 A and 40 volts) No pnp version presently available. Vulnerable to self-oscillation.	High-reliability, low-power output element in both switching and linear supplies. General-purpose power control.
Hybrid Circuit Module NPN Output	Overload protection. Higher power than monolithic devices. Op-amp circuit characteristics. Electrically rugged.	More costly to make than monolithic ICs. Moderate frequency capability. Power capability less than simple transistors.	Regulated power supplies. Public address systems. Motor control. Solenoid drivers. Deflection circuits. Servo amplifiers. Driven inverters. "Workhorse" for 5- to 100-watt loads.

Voltage ratings are sometimes harder to pin down because of recent progress in this area. However, because of the predominance of silicon over the germanium technology, it appears reasonable to now consider any power transistor with a voltage rating of less than 90 volts as a *low-voltage* type. The *medium-voltage* classification would then embrace the range of 90 to 300 volts. Beyond this range are the *high-voltage* transistors that were almost inconceivable not too long ago, but now commonly used in tv sets and for other applications.

In the frequency or speed domain, power transistors capable of switching up to 3 kilohertz without incurring appreciable rise-time and fall-time losses can be considered to be low-frequency devices. From 3 to 15 kilohertz comprises the middle-frequency range, and a

Fig. 3-16. Thermal-cycle rating chart for the RCA RC5258 npn power transistor.

power transistor capable of operating efficiently at higher switching rates qualifies as a high-frequency type. This generally implies that the cutoff frequency, F_T, will be well into the megahertz range. (The cutoff frequency of some older "workhorse" germanium transistors, such as the 2N178, was a mere 10 kilohertz.)

Another helpful guide in transistor selection is the realization that the operating parameters have a strong tendency to be interrelated on a *tradeoff basis*. One can buy germanium power transistors capable of withstanding 500 volts, but their already-poor frequency ratings are further degraded and their low-cost feature is lost. There are two subdivisions of the triple-diffused silicon transistor, one optimized for frequency at the expense of voltage, and the other with high voltage and lower frequency ratings. In Darlington transistors, the high current gain is attained with a sacrifice in speed of response. (It is indeed noteworthy that modern monolithic Darlingtons have been successfully competing with other types in a wide variety of power-control applications.)

Because of the diversity of transistor types and the large number of processing and construction techniques, it should be appreciated that no classifying system can be truly comprehensive. Indeed, if such a system existed, its complexity would defeat its objective. Table 3-2 endeavors to generalize power-transistor use in terms of what now appears to be significant structural differences. An example of a type not listed in the table is the RCA pi-nu multiple epitaxial-base transistor. This transistor structure appears to have successfully blended parameters otherwise tending toward *either* the double-diffusion or ordinary epitaxial-base types. As a consequence, we find this power transistor enjoying widespread application in original-equipment auto ignition circuitry. The RCA 2N6513 exemplifies the ratings of pi-nu devices. Its current rating is 5 amperes and it can withstand 350 volts. Its F_T specification is 3 megahertz and maximum power dissipation is 120 watts. It can readily be appreciated that such a parameter combination optimizes this transistor for service in solid-state ignition systems.

Various transistor structures are shown in Fig. 3-17. Unfortunately, these illustrations are not likely to be as meaningful as the more visible features in vacuum-tube structures. Nonetheless, it is instructive to contemplate the different processing techniques which are responsible for the diversity in the operating parameters of transistors.

Planar construction tends to produce devices with less leakage current than the earlier-developed *mesa* process. In a planar transistor, the junction edges are encapsulated in silicon dioxide "glass." This also tends to reduce noise generation (admittedly, not generally important in power devices) and enhance reliability. *Epitaxial* structures are those in which the manipulation of impurity regions in the semiconductor material is achieved by induced crystalline formation. The thickness and electrical sensitivity of such layers of "grown" material can be precisely controlled. Epitaxial fabrication techniques are particularly useful in lowering collector saturation resistance and in increasing the voltage capability of transistors.

The Darlington transistors have been improved to the extent that they are competitive to the various single-transistor types in a wide range of applications. Where it can be substituted for conventional transistor types, the Darlington transistor often brings about both operating and economic advantages. Because of its high current gain, it can often be directly driven from integrated circuits. When comparing the parameters of all of the power transistors listed in Table 3-2, one finds that the Darlington appears to have the best mix of medium ratings. At the same time, the Darlington lends itself

(A) Single-diffused transistor.

(B) Epitaxial-base transistor.

(C) Double-diffused mesa transistor.

Fig. 3-17. Some examples of

(D) Double-diffused planar transistor.

(E) Triple-diffused transistor.

(F) RCA pi-nu multiple-epitaxial-base transistor.

Courtesy RCA Corp.

silicon transistor structures.

well to optimization of certain parameters. An example is the line of high-voltage Darlingtons available from Delco Electronics. The Delco DTS4075 is a monolithic Darlington with built-in resistors between each base and emitter, as well as an internal speed-up diode to improve switching characteristics. This device has a maximum collector-current rating of 20 amperes. It withstands 350 volts at 1 ampere of collector current.

The synthesized transistor, as exemplified by the National Semiconductor LM195, is a specialized IC. This power device is both inexpensive and cost effective because it incorporates its own reliable protection against current overload or temperature rise. Therefore, one can dispense with external protective devices and circuitry, as well as the time-consuming experimentation often needed to confirm the integrity of protection techniques. Although a relatively low-power device, it is nonetheless a far cry from earlier attempts at IC power devices. At this time, the company is bringing out a "pnp" version. Despite its low cost, the LM195 is certainly among the most sophisticated of the low- and medium-frequency power-control devices, for its monolithic architecture comprises more than 20 equivalent transistors.

While on the subject of nondiscrete power devices, it is appropriate to mention thick-film assemblies and hybrid modules. Although not of monolithic fabrication, such "power packages" resemble the synthesized transistor in the makeup of their internal circuitry—several active devices are connected so as to provide an easy-to-use power-control device. An example is the RCA HC2000H power op-amp. This thick-film/hybrid module is, as its name implies, an operational amplifier with differential input. It can supply peak output currents as high as 7 amperes, and can deliver rms power up to 100 watts. Its bandwidth of 30 kilohertz enables it to render useful service in audio, servo, and deflection amplifiers, and in voltage regulators and driven inverters. Such modules seemingly combine some of the best features of discrete solid-state devices and monolithic ICs.

Double- and triple-diffused power transistors merit consideration when either high frequency or high voltage impose the need for uniquely optimized devices. These structures are not noteworthy for their current-carrying ability or a high SOA. The "specialty" of double diffusion is speed, while that of triple diffusion is high voltage rating. Depending upon doping, geometry, and other aspects of the manufacturer's processing techniques, there can be considerable overlap in application areas for the two types. Also, when using these types, it should be kept in mind that although both switching and rf transistors have the need for speedy response, best results are obtained by ordering devices that are *uniquely specified* for one or the other service. In switching transistors, for example, one is more concerned with a low collector-saturation voltage than in rf types. It should be noted that rf transistors are required to endure unusual stresses resulting from maladjusted tuning and from high vswr. Somewhat different considerations apply to the internal capacitances of the two types, although these capacitances must be low in both instances. In addition, rf types are generally designed for lower base-emitter voltage but greater input current than the switching types. Because of the widespread use of mobile communications, rf transistors are more often intended for 12- or 24-volt operation, whereas switching types may operate at considerably higher voltages. Other subtle differences become more pronounced at high frequencies and emphasize the need to specify the intended service.

Silicon single-diffusion and epitaxial-base transistors are more appropriate for the numerous power-control functions where undue emphasis is not placed upon speed or voltage capability. Today's "workhorse" devices have parameters which were the domain of the "sophisticated" devices of yesterday. Accordingly, both of these structures find application in inverters and switching circuits where good 20- and 30-kilohertz square waves must be produced. Here, some of the older general-purpose germanium transistors would not have been very useful.

At lower speeds, the germanium transistor still has its devotees. If one is prepared to apply more effective heat removal than is required from silicon devices, use of the *germanium* transistor can often prove very satisfactory. It even has the compelling feature of exceptionally low collector-saturation voltage. So, far from being obsolescent, the germanium transistor often finds use in modern power supplies which operate at low voltage and high current. It is also still being used in automobile radios and in economy audio equipment. Although fewer companies make germanium power transistors, a residual demand remains. Because of their high current ratings, tolerance to overload, and low cost, many designers still feel germanium transistors are entirely justified for certain types of motor control, regulators, solenoid drivers, etc.

BASIC TYPES OF SCRs

Many improvements have been incorporated into SCRs since they first became commercially available. These improvements have been targeted at such parameters as triggering reliability, commutation performance, noise immunity, ability to endure and perform in the face of rapidly changing voltage and current, speed, and of course, voltage and current capability. A major dividing line for SCRs pertains to *speed*. The SCRs which are intended for *60-hertz phase-control* service are grouped together, as are those intended for *in-*

verter service. *Phase-control* SCRs often work well at 400 hertz, but at higher frequencies the *inverter* SCR yields optimum performance. Many inverter SCRs appear to be specified for frequencies in the 3- to 10-kilohertz range, but recent developments have pushed the top frequency to the 20- to 30-kilohertz range and there are some available with even higher frequency capability. The terms are somewhat misleading, for it is not the phase control or the inverter operation *per se* that is significant. Rather, it is the frequency at which the SCR must turn on and off. It just happens that most 60-hertz applications utilize phase control, and the higher frequencies are often associated with inverters.

As with transistors, there are different techniques for SCRs. The low-cost "workhorse" for many practical applications has been the *alloyed-diffused* SCR. These SCRs have been very suitable for 60-hertz control applications up to 35 amperes and are available with blocking voltage ratings up to 800 volts. These numbers are not intended as rigid boundaries, but when audio frequency rates and/or greater current or voltage are involved, SCRs constructed via *epitaxial* technology generally merit attention. The SCRs for even more demanding service, where perhaps 2000 volts or more must be blocked together with control of many hundreds of amperes of load current and/or with frequencies of 5 kilohertz or higher, require the most sophisticated of all structures. These devices may combine epitaxial processes, the use of a shunt between gate and cathode, and gold doping. Invariably they contain special gate structures designed to rapidly propagate the turn-on action, as there tends to be a time lapse between the trigger pulse and the anode-cathode response. Large SCRs also must be equipped with effective *thermal hardware*.

APPLYING SCRs AND TRIACS

The most important members of the thyristor family of solid-state power-control devices are SCRs and triacs. They offer efficient control of high power levels and generally involve relatively simple circuits. Nonetheless, successful implementation can be achieved in many instances only by paying attention to a number of important characteristics of these devices. Although the semiconductor manufacturers endeavor to make applications as simple and straightforward as possible, nothing substitutes for knowledge on the part of the user in the quest for trouble-free performance.

In the following discussion, the SCR will be referred to except where it is desired to deal with the unique features of the triac. This is possible because many of the important operating mechanisms are shared by the two devices. As yet, however, triacs lag behind SCRs in voltage, current, and frequency ratings.

Forward Voltage Drop

A commonly encountered notion is that the voltage drop across the anode-cathode terminals of an SCR operating in its conductive state is at least two volts. The reasoning involved is that there are *three* pn junctions in series with a 0.6- or 0.7-volt drop across each one. In actual devices, the measured voltage drop is only that attributable to a single junction plus another few tenths of a volt. For low to medium currents, this voltage drop is in the neighborhood of 0.8 to 1.3 volts, depending upon certain variables such as the device fabrication and the amount of current being handled. The discrepancy stems from the erroneous viewpoint of the device as being *equivalent* to three series-connected diodes. A little reflection will reveal the invalidity of this analogy inasmuch as the "middle diode" is not properly polarized for forward conduction.

This dilemma is resolved by recalling the two-transistor simulated SCR of Fig. 2-24 in Chapter 2. The fact of the matter is that, although the SCR is not *constructed* as a pair of separate transistors, it does indeed *function* as a two-transistor device. The principle current path is through the emitter-base section of a pnp "transistor" and the collector-emitter circuit of an npn "transistor." The regenerative action of the "transistor" pair holds both "transistors" in heavy saturation. The emitter-base section of the pnp "transistor" can only behave as an ordinary diode, contributing about 0.6- to 0.8-volt drop. However, an entire transistor in saturation, in this case the npn "transistor," can have a collector-emitter voltage drop *appreciably less* than that of a diode alone. Thus, the *net* voltage drop from anode to cathode of the SCR is only in the vicinity of that corresponding to some rectifier diodes which have not been processed for an optimally low voltage drop.

Triggering Thyristors

The most common thyristors used in power electronics are SCRs and triacs. They share much in common with respect to triggering techniques since proper triggering is one of the prime operational requirements of both types of thyristors. In the ensuing discussion, particular reference may be made to SCRs, but, unless otherwise stated, similar circumstances will also pertain to triacs.

Gate voltage and gate current for thyristors have relatively wide tolerances from unit to unit. Moreover, the gate trigger requirements are also very temperature sensitive and also vary with the amount of voltage at the anode before conduction. A typical example of gate triggering characteristics for a good quality SCR is shown in Fig. 3-18. One might conclude from these characteristics that a trigger source with plenty of voltage and with adequate current capability should solve any problems stemming from a wide range of gate sensitivities. Although there is considerable truth in such a proposal, it cannot be applied with reckless abandon. The gate is limited in its power-dissipation rating. Also, excess gate power lowers overall efficiency and subtracts from the allowable anode-cathode dissipation. This is particularly true at higher frequencies. Often, the amount of reverse voltage impressed on the gate is also a limiting factor.

Courtesy General Electric Co.

Fig. 3-18. Example of SCR gating characteristics.

Although adequate gate voltage and current might assure the triggering of all thyristors of the same type, it would not solve the problem of divergent control settings for a given load power in phase-control equipment. Ideally, in phase-control applications, the triggering of the thyristor should be governed *only* by the time constant or delay imposed by the trigger circuit. The time of triggering should *not* be influenced by the amplitude of the gate signal. Finally, this brute-force approach would not prevent inadvertent triggering of the most sensitive thyristors by *noise*.

Before designing appropriate trigger circuitry to comply with these variables, the width of the trigger pulses should be considered. It is obvious that if the pulses persist too long after the thyristor has made its intended response, gate dissipation is bound to be un- necessarily high. We can even get into timing difficulties if the trigger signal remains when the thyristor is not intended to be conductive.

Fig. 3-19. Effect of trigger pulse width on minimum required triggering current.

On the other hand, the trigger pulse cannot be made infinitesimally *short* in duration, since the triggering phenomenon requires that a threshold quantity of electrical charge be delivered to the gate. The effect of pulse width for a particular SCR is illustrated in Fig. 3-19. This happens to be a high-frequency SCR, but the curves are quali- tatively similar for all thyristors. In 60-hertz types, the *knee* tends to be in the vicinity of 5 to 10 hertz.

Table 3-3. Solid-State Triggering Devices for Thyristors

CLASS	E-1 CHARACTERISTICS	BASIC CIRCUIT	MAJOR TYPES	V_p PEAK POINT VOLTAGE	I_p (MAX) PEAK POINT CURRENT	I_v (MIN) VALLEY CURRENT	T_{ON} TURN ON TIME	SPECIFICATION NUMBER
UJT Unijunction Transistor			TO-5: 1N489A 5G515 5G516 TO-18: 2N2417A 2N2417B 2N1671A 2N1671B 2N1671C 2N2646 2N2647	Fixed Fraction of Interbase Voltage	12 µA 6 µA 25 µA 6 µA 2 µA 5 µA 2 µA	8 mA 8 mA 8 mA 8 mA 8 mA 4 mA 8 mA	1-2 µsec Typ	60.10 60.10 60.50 60.50 60.50 60.02 60.62
CUJT Complimentary Unijunction Transistor			D5K1 D5K2	Fixed Fraction of Interbase Voltage	5 µA 15 µA	1 mA 1 mA	1 µsec Typ	60.15 60.16
PUT Programmable Unijunction Transistor			D13T1 D13T2	$\dfrac{R_2 V_s}{R_1 + R_2}$	As low as 2 µA As low as .15 µA (A function of R_1, R_2)	70 µA 25 µA (A function of R_1, R_2)	80 nsec Max	60.20

Unidirectional

Type	Characteristic	Circuit	Device Numbers	Voltage			Time	Reference
SUS Silicon Unilateral Switch			TO-18 TO-48 2N4983 2N4987 2N4984 2N4988 2N4985 2N4989 2N4986 2N4990	6.10 V 7.5-9 V 7.5-8.2 V 7-9 V	500 µA 150 µA 300 µA 300 µA	1.5 mA .5 mA 1.0 mA .75 mA	1.0 µsec Max	65.25, 65.26 65.27, 65.28 65.27, 65.25 65.25, 65.26
SCS Silicon Control Switch			3N84	$\dfrac{R_2 V_s}{R_1 + R_2}$ (40V max)	A function of R_1 and R_2	10 mA Max (A function of R_1 and R_2)	1.5 µsec Max	65.18
Bidirectional — SBS Silicon Bilateral Switch			TO-18 TO-98 2N4993 2N4991 2N4992	6.10 V 7.5-9 V	500 µA 120 µA	1.5 mA .5 mA	1.0 µsec Max	65.30, 65.31 65.32
DIAC *Alternate Symbol			ST2	28 V-36 V	200 µA	Very high	1 µsec Typ	175.30

Courtesy General Electric Co.

*Alternate Symbol

Triggering Devices

A number of solid-state triggering devices are available for thyristors. Table 3-3 lists the most frequently used types. It will be noted that all these triggering devices develop sharp pulses when used in appropriate circuits. The peak voltage, peak current, and pulse duration of these devices is compatible with a large number of commonly used thyristors. With very large thyristors, the gate current requirement may necessitate the use of an emitter follower or a small intermediate thyristor as a buffer or driver.

The original diac symbol, depicting a transistor structure with two "emitters," was relevant to the first devices marketed for thyristor triggering. This symbol is now often replaced by the symbol for a "gateless thyristor" (four-layer diode). Whether the actual device referred to as a diac is fabricated with three or four layers, the bilateral switching behavior is much the same.

COMMONLY USED SCR TRIGGERING CIRCUITS

The following circuits illustrate a number of commonly encountered triggering arrangements for SCRs. In the simple circuit of Fig. 3-20, a neon lamp is used as the trigger device. As is the case with many bargain items, this circuit could prove expensive in the long run. One of the disadvantages is the well-known erratic firing characteristics of neon lamps. Aggravating this problem are the unpredictable effects of aging, temperature, light, and electric fields. If this scheme is used, it is recommended that a neon lamp intended for such purpose be selected rather than the "garden-variety" type.

The circuit shown in Fig. 3-21 employs a diac in conjunction with a pulse transformer having a 1:1 turns ratio to obtain full-wave phase control from a pair of back-to-back SCRs. The overall operation simulates that ordinarily obtained with a triac. The circuit in

Fig. 3-20. A neon lamp used as a trigger device.

Fig. 3-21. An SCR triggering circuit using a diac.

Fig. 3-22. A unijunction transistor is used in this SCR trigger circuit.

Fig. 3-23. A half-wave phase-control circuit using a silicon bilateral switch as the trigger device.

Fig. 3-22 utilizes a unijunction transistor fed from rectified ac. The important thing in this arrangement is that no filter capacitor should be used with the dc supply because it is necessary that the operating voltage be periodically returned to zero.

In the half-wave phase-control circuit of Fig. 3-23 the trigger device is a silicon bilateral switch. This is an IC with the simulated characteristics of the diac. The structure involves neither the four-layer fabrication nor the double-emitter transistor commonly employed in bilateral switches. In this case, the circuit function is *synthesized.* Another difference from the diac is the availability of one or more *gates,* which may be optionally used for various logic techniques. In this application the use of the gate always causes the capacitor to commence its charge cycle from the same near-zero voltage. This makes the control hysteresis-free and prevents the irritating "snap-on" effect which plagues many "economy-type" lamp dimmers. The diode protects the SCR gate from reverse biasing.

A somewhat similar trigger scheme is shown in circuit Fig. 3-24 which uses a silicon unilateral switch. A unique feature of this circuit is that the power can be controlled in either a dc or an ac load. The dc output voltage from the bridge rectifier should not be filtered, however, because the commutation of the SCR is achieved by the periodical return of this voltage to zero.

The trigger circuit in Fig. 3-25 makes use of an optocoupler. The salient feature of this triggering technique is the electrical isolation obtained between control circuitry and the load. Optocouplers with

Fig. 3-24. A triggering circuit using a silicon unilateral switch.

1500-volt isolation between the input LED and the light-sensitive output device are commonly available. Higher voltage units are also marketed. Triggering via optocouplers is especially popular in solid-state relays. This technique also merits consideration in systems where it is necessary to prevent false triggering from noise or transients.

Any mention of optoelectronic techniques naturally suggests the light-actuated SCR. These devices inherently provide electrical isolation between control or trigger circuits and the load. Fig. 3-26

140

INPUT →

Fig. 3-25. An optocoupler used to
trigger an SCR.

depicts an arrangement of light-actuated SCRs for full-wave control
of load power from a high-voltage ac line. All of the gates are illuminated simultaneously. Sometimes this is achieved with the use
of fiber optics. As shown, the circuit can be used as a switch by maintaining the gate "signal" as long as it is desired to energize the load.
Such triggering does not cause gate dissipation as do all electrical
gate-triggering methods.

There is more than meets the eye in the partial circuit of Fig.
3-27 using the General Electric PA424 module which was one of the
first successful zero-crossing voltage detectors. As shown, the circuit
delivers full-wave power to the two load SCRs. When this circuit
is switched on, it only applies load current when the voltage

Fig. 3-26. Full-wave control of load power by light-activated SCRs.

141

Fig. 3-27. A trigger circuit using a zero-crossing voltage detector.

sine wave is crossing its zero level. Such a scheme generates negligible rfi, and, therefore, filters are generally not needed. Control of
load power is usually acomplished by burst modulation—varying
numbers of *complete* ac cycles are applied to the load. Such bursts
are interspaced by "dead times" in which no current is available for
the load. The complete circuitry for doing this is not shown since the
present focus is on the triggering technique itself. A number of ICs
are now available for acomplishing zero-voltage control and for
facilitating feedback arrangements so that the desired control or
regulation can be realized. This triggering method spares thyristors
from considerable stress since both dv/dt and di/dt are much less
than is likely to be encountered in phase-control circuits.

The phase-control triggering circuit of Fig. 3-28 uses a programmable unijunction transistor (PUT). A noteworthy feature of this
arrangement is that the firing point of the PUT can be adjusted by
varying the resistance associated with its anode gate. In essence, we
have a "variable diac."

Fig. 3-28. A phase-control triggering circuit using a programmable
unijunction transistor.

142

Fig. 3-29. Circuit providing alternate triggering of two SCRs.

The triggering circuit of Fig. 3-29 provides essentially the same kind of load power control as does the arrangement shown in Fig. 3-21. An advantage of the configuration of Fig. 3-29 is that total electrical isolation is obtained between the load and the trigger circuitry. However, a three-winding pulse transformer is required.

Fig. 3-30. An SCR triggering circuit using a bipolar transistor.

Fig. 3-31. Trigger circuit using a programmable unijunction transistor as a pulse sharpener.

The transistor circuit shown in Fig. 3-30 can be tailored to operate in either of two modes. It can perform as a linear emitter follower. Its chief usefulness, however, results when the transistor is driven into heavy saturation. It then acts as a *switch* which abruptly transfers the charge accumulated in capacitor C1 to the gate circuit of the SCR.

The trigger circuit in Fig. 3-31 is best termed a *pulse sharpener* because it will transform a slow rising pulse into one having a rise time of 50 to 100 nanoseconds. When a relatively slow positive-going pulse is applied to this circuit, the 0.2-microfarad capacitor commences charging through the 1N4154 diode. As long as the charging process continues, the D13T programmable unijunction transistor (PUT) will not fire because the anode gate will be *more* positive than the anode by at least the voltage drop of the diode. However, when the pulse attains its peak value and starts its decline, the anode gate will become *less* positive than the anode because of the retention in the capacitor of the near-peak voltage level. This condition, enhanced by the fact that the diode becomes reverse-biased, enables the PUT to fire. As mentioned, the resultant SCR trigger pulse then has an extremely fast rise time.

The arrangement illustrated in Fig. 3-32 has two implementations. The predriver SCR may either be external to the main SCR, or it

Fig. 3-32. Amplifying gate SCR used to trigger main SCR.

may be integrally fabricated with the main SCR. In the latter instance, the composite structure is known as an *amplifying gate SCR*. Such a device often enables the circuit designer to incorporate a high-power SCR into a system with the use of the same trigger circuits ordinarily employed at lower power levels.

When the SCR has to turn on fast-rising load currents, reliable triggering often demands a rectangular pulse with very fast rise time. A better approach to this requirement than any of the previous circuits is the circuit shown in Fig. 3-33. Here, a UJT relaxation oscillator is followed by a cascade arrangement of four bipolar transistors that takes advantage of the cumulative storage time to lengthen the pulse. The net result is a fast-rising rectangular trigger pulse about 10 microseconds in duration that has a 20-volt amplitude with a source impedance of 20 ohms. It can be produced at a rate of up to 20 kilohertz.

The triggering circuit shown in Fig. 3-34 provides sustained excitation of the SCR gate, but minimizes power dissipation through a

Fig. 3-33. An SCR triggering circuit using a cascade of four bipolar transistors to lengthen pulse.

sequence of narrow pulses rather than a prolonged rectangular wave. Such sustained triggering is often required when a highly inductive load is being controlled. Another feature of this arrangement is that two pulse trains displaced 180 degrees in low-frequency time are generated. (One sequence of high repetition rate pulses is on while the other is off, and vice versa.) Such an operational mode is ideal for inverters. The alternations are brought about through the action of the multivibrator consisting of transistors Q1 and Q2, and associated circuitry. When transistor Q1 is in its conductive state, the re-

Fig. 3-34. Trigger circuit providing a sequence of narrow pulses to minimize power dissipation.

145

laxation oscillator including UJT Q3 is prevented from producing the high repetition rate trigger pulses. Similar inhibitory action occurs during the subsequent half of the multivibrator cycle when UJT Q4 is prevented from oscillating because of the conductive state of transistor Q2.

A somewhat similar, but simpler, triggering circuit for inverters and push-pull or bridge-type power-control systems is depicted in Fig. 3-35. The trigger pulses occur in flip-flop fashion, enabling one SCR of a pair to be triggered when the other SCR is off. Single sharp pulses are generated rather than a burst. The circuit consists

Fig. 3-35. Triggering circuit for inverters and push-pull or bridge-type power-control systems.

of two PUT relaxation oscillators arranged so that a trigger pulse generated by one of them commutates the other oscillator. In this way, each oscillator is allowed to generate only one pulse at a time. Commutation takes place via capacitor C_T. Adjustment of the repetition rate is provided by potentiometer R1. Symmetrical spacing of the pulses can be attained by adjustment potentiometer R2.

A series-connected arrangement of SCRs, which is able to hold off many thousands of volts, can be utilized in high-voltage inverters and switches. It is, however, necessary that all of the gates be triggered simultaneously. It is also very desirable that the gate circuits be electrically isolated from the high-voltage system. A convenient way of meeting these objectives is with a triggering scheme such as is shown in Fig. 3-36. The transformer has a separate secondary winding for each SCR and must be designed to withstand the maximum voltage of the load circuit with an adequate safety margin.

The circuit shown in Fig. 3-37 is a zero-point switch. It uses discrete components to accomplish a triggering function similar to that

146

Fig. 3-36. Series-connected SCRs used in triggering scheme for high-voltage inverters.

of the IC arrangement in Fig. 3-27. As is often the case, there is more here than meets the eye. The circuit is conveniently analyzed by initially assuming capacitor C2 is *not* present in the circuit.

Suppose that switch S1 is turned on and the circuit is energized during the positive alternation of the ac line voltage. There will be *no* action of any consequence because diode D1 is reverse biased. Commencing with the negative alternation of the ac line, capacitor C1 will charge. The maximum voltage it can develop will depend upon the peak voltage and upon the ratio of resistors R1 and R2. This maximum voltage turns out to be about −40 volts. After the negative alternation has reached its peak, capacitor C1 will *remain*

Courtesy Motorola Semiconductor Products, Inc.

Fig. 3-37. A zero-point switch using discrete components.

147

charged at approximately −40 volts because diode D2 will then be reverse biased and diac D4 does not have sufficient voltage to fire. However, the voltage appearing across diac D4 *increases* as the negative alternation of the line voltage *declines*. (The line voltage and the voltage stored in capacitor C1 are analogous to two batteries connected in *series-bucking*.) When the line voltage decreases to about 10 volts, the diac will "see" its triggering potential which is about 30 volts. Note that the SCR gate receives its trigger *before* it is ready to respond—that is, the SCR is triggered while its anode is still *negative* with respect to its cathode. Because of the amount of charge stored in capacitor C2, the SCR gate signal will endure through the zero crossing of the ac line voltage. This action ensures true zero-cross triggering. (Some circuits provide the gate signal slightly *after* zero-crossing and can thereby generate a considerable amount of rfi.)

Now let us consider the effect of capacitor C2 in more detail. The circuit action is the same up to the point where capacitor C1 begins to discharge through the diac and the gate circuit of the SCR. But now, the instantaneous discharge path will be through D3, D4, R3, the gate of SCR Q1, and capacitor C2. Capacitor C2 becomes quickly charged, thereby reducing the diac voltage and causing it to turn off. Capacitor C2 will then discharge through resistors R1 and R2 until the net voltage across the diac again becomes sufficient to turn it on. This again replenishes the charge in capacitor C2, which again reduces the net voltage across the diac so that it reverts to its blocking state. The time constants are such that *during* an interval embraced by a short time *before* zero-crossing and a short time *after* zero-crossing, a *series* of gate pulses are delivered to the SCR. Thus, the addition of capacitor C2 provides the zero-crossing function with less gate dissipation in the SCR than prevails with a single, prolonged, trigger pulse.

TRIGGERING TRIACS

The triggering of triacs is essentially similar to that of SCRs except that *two* trigger pulses per cycle are required. The situation is similar to that of full-wave SCR circuits. However, the circuit implementation of triacs is simpler than that of full-wave SCR arrangements because there is only *one* gate.

In referring to the electrodes of the triac, it would not be relevant to call them an *anode* and a *cathode*. Rather, the connections having to do with the load current are known as *main terminal 1* and *main terminal 2*. All voltages at the gate and main terminal 2 are referred to main terminal 1.

The triac will trigger with *either* a positive or a negative gate signal regardless of whether main terminal 2 is positive or negative. It might appear that this would complicate the circuit implementations because of inadvertent triggerings. It turns out, however, that most triacs have greatest gate sensitivity when both the gate and main terminal 2 are positive or when both the gate and main terminal 2 are both negative. Other polarity combinations for the gate and main terminal 2 are less likely to result in triggering. On the other hand, triacs are available with various gate-polarity sensitivities. It is a matter of selecting the triac which is appropriate for the particular control application.

A simple switching circuit for a triac is shown in Fig. 3-38A. The triac, in response to a low-level gate signal, provides considerable power to the load. In essence, this simple arrangement is a solid-

(A) Simple switch control of a triac.

(B) Basic diac circuit for phase control of load power.

(C) Full-wave phase control with unijunction transistor.

(D) Asymmetrical trigger device for hysteresis-free control.

Fig. 3-38. Various devices and techniques for triggering triacs.

state relay. The main advantages over its electromechanical counterpart is that there is no arcing, and there is virtually no maintenance required. Triacs are usually less costly than quality relay-type devices for handling high currents and they generate less rfi and emi. In this circuit it can be seen that the gate is positive when main terminal 2 is positive. Also, both electrodes are negative at the same time and there is no opportunity for other polarity combinations. Under these conditions, the triac remains turned on through the entire 360 degrees of the power-line cycle. *Initially,* however, the triac may be triggered at *any* point in either the positive or the negative alternation. Therefore, *some* rfi and emi is likely to be produced when the read switch is first closed.

Fig. 3-38B is representative of the basic diac-triac phase-control system. Full-wave control of load current is the prominent feature of such an arrangement. Functionally, it is equivalent to a phase-control circuit using a pair of back-to-back SCRs. The fact that the single gate of the triac suffices for control of both power-line alternations obviously brings about simplification. The bilateral triggering characteristics of the diac conveniently complies with the gate characteristics of the triac.

Another commonly used trigger circuit for triacs is shown in Fig. 3-38C. Here, an unjunction relaxation oscillator is used in much the same manner that it is often encountered in SCR triggering systems. Of course, *two* trigger pulses per cycle are utilized to impart full-wave control of the triac. It should be realized that the triggering mode experienced by the triac is *not* the same as in the diac circuit of Fig. 3-38B. When main terminal 2 of the triac is positive, *this* circuit delivers a negative signal to the gate, as contrasted to the positive gate signal produced by the diac circuit. Both circuits trigger the gate negatively when main terminal 2 of the triac is negative. For best results in either circuit, it is generally wise to *specify* the triggering mode for the triac selected.

The phase-control scheme depicted in Fig. 3-38D is schematically similar to the diac control circuit of Fig. 3-38B. The triggering device is a General Electric ST4 asymmetrical trigger IC. Although *bilateral* like the diac, it has nonsymmetrical breakdown characteristics. The breakdown characteristics of a diac and the ST4 are shown in Fig. 3-39. It might seem that any deviation from the symmetrical-breakdown-voltage feature of the diac could lead to undesirable performance. This is not the case, however, since the symmetrical characteristic of the diac control circuit is objectionable in certain applications, such as in light dimming. When the diac circuit is used, the lamp load cannot be gradually turned on. Rather, the lamp will suddenly glow at moderate intensity as one attempts to increase light intensity from zero. This is because the timing capacitor is

(A) Diac.　　　　　　　　　(B) ST4 trigger device.

Courtesy General Electric Co.

Fig. 3-39. Breakdown characteristics of a diac and the ST4 trigger device.

initially uncharged. As the resistance value of control R1 in Fig. 3-38B is slowly decreased, it would be natural to suppose that very small triac conduction angles could be gradually brought into play, and that a load could be smoothly adjusted from zero light intensity to brighter levels. However, such is not the case. Let us see why.

Referring to the waveform diagram of Fig. 3-40, which pertains to the diac circuit in Fig. 3-38B, assume that the resistance of phase control R1 is slowly being decreased from its initial maximum-resistance setting. The capacitor voltage will increase until the diac (and therefore, the triac) triggers at point A_1. The corresponding triac conduction angle is θ_1. Everything would be fine if the next triggering point were A_2, since the triac would then have a steady-state conduction angle of θ_1 and the lamp load would receive very low current. One could then gradually adjust R1 for progressively more lamp current. Unfortunately, however, the triggering advent of

Courtesy General Electric Co.

Fig. 3-40. Waveforms showing diac hysteresis effect.

151

A_1 depletes the charge in the capacitor, thereby *lowering* its stored voltage. The capacitor voltage will not, therefore, follow the ideal path shown by the dashed line. So the diac will next trigger at point B rather than at point A_2, which is the symmetrical counterpart of point A_1.

Although point B corresponds to the same diac triggering voltage as does point A_2, point B occurs at an *earlier* time than point A_2. This results in an abruptly large triac conductance angle θ_2. Thereafter, all subsequent triggerings take place so that conductance angle θ_2 is maintained. That is why the lamp suddenly glows at a moderate brightness as resistance R1 is decreased. It is true that the resistance of R1 can then be adjusted for low light intensities. Obviously, such an adjustment procedure is a nuisance. This hysteresis is appropriately known as the "snap-back" effect in light dimmers. Paradoxically, the use of the symmetrical triggering device, the diac, does *not* lead to symmetrical operation during start-up.

The nuisance value of hysteresis can be even more irritating. Suppose that start-up has been accomplished and the adjustment control is then turned back to produce a low light intensity. If there is a momentary decrease in line voltage, as might occur during turn-on of a heavy appliance, the light could be extinguished and remain so. To turn it on again, one would have to again advance and then retard the adjustment control.

With the ST4 asymmetrical trigger device, triggering would not occur at point B. Instead, a *higher* voltage would have to be reached, but this consumes more time, thereby delaying the firing time of the triac. When the triac does finally fire, the conduction angle, θ_2, is much smaller than indicated in Fig. 3-40. In fact, the design of the triggering voltages of the trigger device is such that conduction angles θ_1 and θ_2 are very nearly the same. Thus, hysteresis is virtually eliminated.

Hysteresis can also be greatly reduced by the use of a *double-time-constant triggering network*. Such a circuit is shown in Fig. 3-41. Capacitor C2 behaves as a charge reservoir and quickly replenishes much of the charge lost by capacitor C1 when triggering occurs. Fig. 3-42 illustrates that the ideal and actual voltage waves

Fig. 3-41. Triac-diac lamp-control circuit with double-time-constant triggering network.

Courtesy RCA Corp.

Courtesy RCA Corp.

Fig. 3-42. Waveform of triggering actions in double-time-constant trigger circuit.

of capacitor C1 are nearly coincident. Thus, the "snap-back" effect is minimal if not negligible.

The circuit shown in Fig. 3-43 uses a zero-voltage switch, the RCA CA3059, to gate the triac. This versatile IC provides zero-voltage-crossing triggers as indicated in Fig. 3-44. Many different arrangements can be implemented with this device so that burst modulation, feedback control, and various logic and inhibit functions can be realized. Not only is there no rfi or emi generated by this switching

Courtesy RCA Corp.

Fig. 3-43. Triac triggering circuit using the RCA CA3059 zero-voltage switch.

FREQ. (Hz)	T (mS)	t (μS)
60	8.3	100
400	1.25	12

Courtesy RCA Corp.

Fig. 3-44. Waveform diagram showing trigger pulses of RCA CA3059 zero-voltage switch.

technique, but the triac is subjected to less electrical stress than generally prevails in phase-control triggering.

THINGS TO KEEP IN MIND ABOUT THYRISTORS

A number of factors merit consideration in the interest of successful performance from thyristors. First, it is always wise to try to provide adequate heat removal. In addition to avoiding excessive junction temperatures, another reason for keeping operating temperature low is that all parameters become degraded with rising temperature. It is not an easy task to maintain proper operating conditions if temperature swings are wide. For example, important parameters of triggering voltage and current have sloppy tolerances at best, but are worsened with temperature change. Other matters being equal, it is easier to avoid such malperformances as unreliable triggering, false triggering, and commutation problems if the case temperature of the thyristor is kept relatively constant. Of course, when economy is important, the main consideration then is to keep junction temperatures within maximum rating. However, be prepared to cope with challenging circuit demands from your hot thyristors.

Surge Current

A common mistake is to select a thyristor with a current rating apparently properly matched to the steady-state requirements of the load. The thinking is right, but does not go far enough. For example,

Table 3-4. Surge-Current Characteristics of Standard Lamps

Wattage	Rated Voltage	Type	Amps. Steady State Rated Voltage	Hot/Cold Resistance Ratio	Theoretical Peak Inrush (170 V pk) (Amps)	Rated (Lumens/Watt)	Heating Time to 90% Lumens (sec)	Life Rated Hours Average
6	120	Vacuum	0.050	12.4	0.88	7.4	.04	1500
25	120	Vacuum	0.21	13.5	4.05	10.6	.10	1000
60	120	Gas Filled	0.50	13.9	9.70	14.0	.10	1000
100	120	Gas Filled	0.83	14.3	17.3	17.5	.13	750
100 (proj)	120	Gas Filled	0.87	15.5	19.4	19.5	.16	50
200	120	Gas Filled	1.67	16.0	40.5	18.4	.22	750
300	120	Gas Filled	2.50	15.8	55.0	19.2	.27	1000
500	120	Gas Filled	4.17	16.4	97.0	21.0	.38	1000
1000	120	Gas Filled	8.3	16.9	198.0	23.3	.67	1000
1000 (proj)	115	Gas Filled	8.7	18.0	221.0	28.0	.85	50

the *cold* resistance of lamp loads may easily be ten or fifteen times less than the hot resistance. See Table 3-4. For a brief time, the thyristor may therefore be called upon to endure gross overload. When evaluating current capability, it is important to base selection upon surge-current ratings as well as the steady-state specifications. A typical surge-current rating curve is shown in Fig. 3-45. Motors are also notorious for their initially high current demands and for their inrush period that may last much longer than that of lamps. It is both an art and a science to make the most cost-effective provisions for surge-current requirements. However, most specification sheets provide relevant data for this important consideration. Keep in mind

Courtesy RCA Corp.

Fig. 3-45. Typical surge-current rating curve for power transistors.

that a thyristor with a hot, but safe, junction temperature under *steady-state* operation will be more susceptible to destruction from a turn-on surge than will be a device provided with more effective heat removal which lowers steady-state junction temperatures.

Inductive Loads

A peculiar commutation difficulty for triacs arises from the effect of inductive loads. Consider the triac circuit shown in Fig. 3-46. Suppose load-current control is attempted through time-delayed gate pulses. This is the technique commonly employed in phase-control thyristor circuits used with essentially resistive loads such as heaters and lamps. With an *inductive* load, however, the load current

Fig. 3-46. Waveform diagram for triac circuit showing how commutation can be lost with inductive load.

will lag the line voltage as illustrated by the waveform diagram in Fig. 3-46. As a consequence, when the load current approaches zero and the triac turns off, it will suddenly be impressed with a line voltage that is not zero. Even though the value of this voltage will obviously be far below that corresponding to the maximum blocking voltage of the triac, its *rate of application* (dv/dt), will be exceedingly great. Accordingly, almost instantaneously turn-on will occur and the commutating ability of the triac will be lost.

The usual remedy is the addition of the RC *snubber network* (shown by dashed lines) which limits dv/dt to about one volt per microsecond when C_1 is 0.1 microfarad and R_1 is 100 ohms in 60-hertz circuits. However, a triac should be chosen that has a dv/dt capability of at least 2 volts per microsecond. The purpose of the resistor is to damp any tendency toward ringing of the LC circuit. This would aggravate rather than alleviate commutation problems.

Another commutation problem can be experienced with inductive loads. Because the inductance initially slows down the rise rate of the load current, the thyristor may not have attained holding current by the time the gate trigger pulse is over. In this case, there will be no sustained turn-on of load current and, for practical purposes, it will be as if the thyristor has not been triggered. This situation can be alleviated by prolonged trigger pulses, or by a series of narrow gate pulses occurring at a high repetitive rate. Sometimes one can experiment with the resistance and capacitance values of the snubber network in order to bring about a situation where the *discharge* of the capacitor through the triggered thyristor will provide sufficient holding until the rising load current eventually "takes over."

The application of the snubber network to full-wave SCR circuits is similar to that described for triacs. However, such back-to-back SCRs generally have better dv/dt capability to begin with. Even half-wave SCR circuits often have snubber networks. Here the protection may not be needed so much to prevent the effect of inductive lag, but to slow down transients from the ac line, motors, and other sources.

Blocking and Reverse Voltage

A thyristor must have voltage ratings compatible with the circuit in which it operates. Otherwise, it will not be able to *block* the applied voltage in the absence of gate trigger pulses. With an SCR, one must make certain that the voltage is never high enough to cause reverse avalanche breakdown. Fortunately, this requirement is usually satisfied if the forward blocking state is safely maintained.

Rate of Voltage Application

Compliance with voltage conditions just stated above is only a beginning—one must also have some notion of the *rate of applied voltage change* (dv/dt) across the main terminals of the thyristor. If this rate is too fast, the thyristor will turn on without a gate signal. This is because the internal junction capacitances will pass a sufficient current to make the thyristor "think" it has received a gate signal. The mechanism is a simple manifestation of the fact that any capacitance passes current in an amount directly proportional to the rate of change of the impressed voltage. The dv/dt ratings are important under two conditions. First, *applied* dv/dt pertains to the simple application of main-terminal voltage to the thyristor. Any thyristor will ultimately self-trigger when such voltage is initially applied too abruptly. *Reapplied* dv/dt is a similar concept but relates to the situation where the thyristor has just been turned off. The significant point is that a certain time must lapse for the thyristor to regain its blocking ability. Such reapplied dv/dt, also known as *commutating* dv/dt, varies with the frequency and the magnitude of the load-current pulses.

It is important to keep in mind that the dv/dt capabilities of a thyristor can be exceeded by voltages appreciably *smaller* in magnitude than the forward blocking voltage. Indeed, if the rate of change for the voltage is sufficiently great, a voltage with relatively low magnitude can cause false triggering or prevent proper commutation. Such disturbances are often superimposed on the ac power line in the form of transients which, ironically, may originate in *other* thyristor equipment.

Dealing with Transients

Remedial measures generally consist of one or more of three basic techniques. First, we should select a thyristor with good dv/dt ratings. Second, sharp pulses can be slowed down by RC networks known as *snubbers*. And third, various voltage-clipping or energy-absorbing devices can be used. The most applicable of these devices are zener diodes and the General Electric MOV varistor. This device is fabricated from polycrystalline zinc oxide. Its electrical behavior is similar to that of a back-to-back zener diode arrangement—that is it exhibits bilateral "crowbar" characteristics similar to the avalanche breakdowns of zener diodes. Its prominent features are its high energy-absorption capability and its speedy response.

Rate of Load-Current Application

We have just discussed thyristor malperformance as a result of slowing down the rate of load-current rise following triggering. It is also true that trouble is encountered if this rate, symbolically known as di/dt, is too high. The reason for this is that the turn-on process for a thyristor is not instantaneous. Rather, some time is required for the current to spread over the cross-sectional area of the semiconductor material. In the thyristor, the demand for high load current before turn-on has propagated throughout the junction area in the vicinity of the gate can be destructive, for the current *density* then exceeds safe values. This can result in device deterioration, or in quick catastrophic destruction. Although the internal mechanisms and circuit conditions are not quite the same as in secondary breakdown of a transistor, the destructive phenomena in the two devices has a common characteristic. In both instances, localized "hot spots" are produced that exceed average junction temperature.

A di/dt that is too large will not be experienced in simple inductive circuits. Indeed, stray inductance in connecting leads often suffices as protection. However, high-frequency circuits working into essentially resistive loads, particularly at high power levels, often require di/dt consideration. Another source of excessive di/dt can be the snubber network. It has already been pointed out that the discharge of the capacitor through the thyristor can sometimes be advantageously used to *speed up* di/dt in inductive circuits where triggering difficulties are experienced. However, the capacitor discharge rate must not be too great, which is another reason for the series resistor. Commutation circuits also have to be monitored closely to be sure that they do not produce excessive current demand.

In addition to taking appropriate circuit precautions, it is necessary to *select* thyristors with di/dt ratings suitable for the particular application. As with power transistors, different device fabrications are available for optimization of operating parameters. Also, the law of trade-offs applies in analogous fashion. An example of this is found in the rule-of-thumb relationship for 60-hertz workhorse SCRs:

$$I_{T(AVG)} = \frac{100\sqrt{t_q}}{V_{RM}}$$

where,

$I_{T(AVG)}$ is the rated average current in amperes,
t_q is the rated turn-off time in microseconds,
V_{RM} is the rated blocking voltage in kilovolts.

A NEW ERA IN SOLID-STATE POWER

From out brief excursion into heat theory in the previous chapter, it should be evident that the overwhelmingly important factor in establishing power ratings for solid-state devices is the efficiency with which heat can be transferred away from a device. This inspires an unending quest for lower overall thermal resistance between the pn junction, or other region where heat is developed, and the ambient environment. Unfortunately, progress becomes very slow once we have resorted to massive copper, aluminum, or beryllium-oxide structural elements, good craftsmanship, and, of course, large heat sinks. Semiconductor makers have been obsessed with the possibility that some material might be discovered which would conduct heat tens or even hundreds of times more effectively than ordinary substances. In essence, they have found it! The two solid-state devices shown in Fig. 3-47 both have 400-ampere ratings. The disk-like diode in the foreground and the required heat sink behind it represent what has up to now been considered a highly efficient heat-removal technique for convection cooling. The tiny device at the right is also a 400-ampere diode, but it requires no additional thermal hardware! This dramatic comparison represents a technological breakthrough that has been brought about by RCA with their new line of *transcalent* devices, which includes diodes, transistors, and SCRs.

These new solid-state power devices make use of a so-called *heat pipe* to transfer the heat from the device to the heat sink. The inordinately small heat sink will be shown to also be the consequence of the heat pipe. Heat pipes make use of a working fluid (often water) to transfer heat from one region to another. Its use is based upon the fact that fluids absorb a tremendous quantity of heat when they

Fig. 3-47. Conventional 400-ampere diode and required heat sink (left) compared with new 400-ampere diode developed by RCA.

Fig. 3-48. Cutaway view of a experimental heat pipe.

evaporate and then, when they condense again, they liberate this heat to the environment. This suggests the nature of the heat pipe. It is a closed system in which the working fluid is converted to its vapor state by the heat from a heat source such as a power transistor. The vapor then expands to a cooler region and liquefies again, liberating its heat in the process. Finally, it is pumped back to the heat source, whereupon the cycle is repeated.

The cutaway view of an experimental heat pipe shown in Fig. 3-48 provides useful insights into the practical aspects of this device.

161

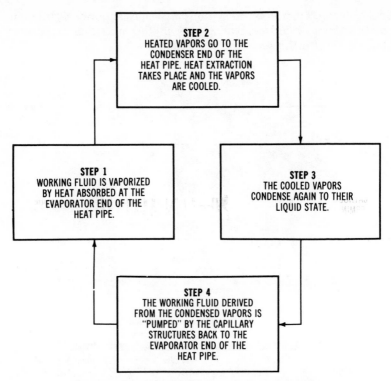

Fig. 3-49. The operational sequence in the heat pipe.

The salient feature of the heat pipe is that no mechanical pump is needed to keep the working cycle going. Rather, the "pumping" of the condensed fluid back to the heat source is accomplished by a *capillary structure,* or wick. Such a structure can be woven cloth, fiberglass, wire screen, porous metal or ceramic, or narrow grooves cut longitudinally into the inner surface of the pipe. An example of capillary action is the continuous rise of molten wax in a candle wick. The block diagram of Fig. 3-49 illustrates the sequence of events within the heat pipe.

In the RCA transcalent power devices, heat pipes are bonded to both sides of the silicon wafer. Fig. 3-50 depicts such an arrangement for an SCR. Essentially, the same fabrication technique is also used for diodes and in power transistors. Such structures place power semiconductor ratings in an entirely new range. Now these transcalent devices are considered exotic and their high cost is justifiable only in military and in space applications. The decline in price usually experienced by such new products should soon set in, and these high-performance devices will merit consideration for service in welding, electroplating, motor control, transport vehicles, power distribution, metal refining, induction heating, and in other industrial areas where the control of high power levels is involved.

AUX. CATHODE LEAD

GATE LEAD

SILICON WAFER

HEAT PIPE

CATHODE TERMINAL

ANODE TERMINAL

WICK

CERAMIC TO METAL SEAL ENVELOPE

Courtesy RCA Electro Optics & Devices

Fig. 3-50. Integrated structure consisting of heat pipes bonded directly to the SCR wafer.

4

Circuit Applications Using Power Transistors and Power ICs

HEADLIGHT BEAM CONTROLLER

The interesting application of solid-state circuitry shown in Fig. 4-1A automatically lowers the high headlight beams of an automobile in response to the headlights of an oncoming car. The idea is not new, being an adaptation of options offered to buyers of Cadillac and other GM automobiles in the fifties and sixties. However, those versions of this scheme used vacuum-tube circuitry as illustrated in Fig. 4-1B. It will be seen that the vacuum-tube model made use of a phototube, a 7851 preamplifier tube, and a 12K5 amplifier tube. The solid-state version is a better performer, is more reliable, and surprisingly, utilizes fewer components in a simpler circuit.

When the phototransistor receives incipient light from an oncoming car, its conductivity is increased and it consumes increased collector current through the 22-megohm load resistance. (As is conventionally done in many phototransistor circuits, the base is left electrically unconnected—one might consider the *light* as the source of base "bias.") The increase in collector current of the phototransistor is then amplified in the D32S4/D38S4 first amplifier stage. Additional current gain is then provided by the D41K3 relay driving stage and the actuation of the relay shifts the headlight circuit from the high beam to the low beam. Two factors contribute greatly to the sensitivity of this circuit. First, the use of a phototransistor rather than a photodiode essentially dispenses with an additional amplifier stage. Second, the D41K3 device is actually a power Darlington with an exceptionally

164

RELAY: 12V, 0.3A COIL; 20A, FORM C, CONTACTS
150A COLD FILAMENT SURGE, RATING.

(A) Solid-state circuit.

(B) Original vacuum-tube circuit.

Courtesy General Electric Co.

Fig. 4-1. Automatic beam controller for automobile headlights.

high current gain as shown by the graph in Fig. 4-2. Thus, the
power output stage, rather than being a trivial contributor to overall
sensitivity, can be said to be a major factor. Also, the elimination of

Fig. 4-2. Representative curve showing the high current gain of the D41K3.

discrete amplifier stages in such a direct-coupled system helps attain stability and performance reliability.

Of course, the alert experimenter may wish to use all solid-state components by replacing the electromechanical relay with a thyristor or a power transistor. A problem here is that the headlights consume about 16 amperes when the filaments are hot, but cold filaments require a 150-ampere surge. Also, semiconductor "contacts" will be less perfect than physical contacts insofar as concerns voltage drop. (Balancing this is the practical fact that the resistance of physical contacts increases with age and use. Also, the resistance of metallic contacts increases with temperature—the opposite is true of a semiconductor resistance.)

Inspection of the circuit in Fig. 4-1A will reveal that three cascaded common-emitter stages are involved, counting the phototransistor as the first stage. However, some liberties have been taken with regard to the ordinary configuration of such a circuit. Specifically, the positive feedback path provided by the 0.1-μF capacitor endows the circuit with hysteresis. Once activated, the circuit remains in that state for a time; that is, the low beam will remain energized even when the light impinging upon the phototransistor disappears. A little thought will indicate that this is necessary in order to prevent objectionable hunting or flashing of the headlights. The DT230F diode connected across the relay coil protects the output transistor from the

166

inductive transients generated when the relay-coil current is abruptly interrupted.

It is recommended that a lens with a minimum diameter of one inch be positioned in front of the phototransistor to provide a "viewing" angle of approximately ten degrees.

A POWER IC INTERCOM

The simple circuit shown in Fig. 4-3 is an intercom system designed around an LM384 IC audio power amplifier. Outputs of up to 5 watts can be obtained under the most favorable conditions which involve a 26-volt supply, an 8-ohm speaker, and a clip-on heat sink. However, outputs of several watts are readily obtained with a 20-volt

Courtesy National Semiconductor Corp.

Fig. 4-3. Intercom system utilizing an IC audio power amplifier.

power supply and with heat removal provided by several square inches of copper foil on a printed-circuit board. Satisfactory operation at somewhat reduced output is obtained for either 4- or 16-ohm speakers. The quiescent current drain from the power supply is low because the output stage operates very nearly in Class B. The IC is designed to have a fixed voltage gain of 50, which is why the 25:1 microphone transformer is necessary.

The LM384 has internal thermal limiting and the output is short-circuit proof. It is in a 14-pin dual inline package with six of the pins serving to improve heat transfer to a printed-circuit board. (These six pins are at electrical-ground potential.)

If the master and slave stations are poorly isolated, acoustic feedback may occur with the volume control, R_v, at its maximum setting. If such howling or squealing is experienced, it can often be remedied by reversing either the primary or secondary connections of the microphone transformer. The physical orientation of the speakers is also an influencing factor in such feedback.

ELECTRONIC SIREN

The siren circuit illustrated in Fig. 4-4 is simpler than it appears inasmuch as the transistors and the "op-amp" are integral to a single IC package. It will be noted that the transistor terminals are numbered along with the amplifier symbol. Useful circuits can be con-

Courtesy National Semiconductor Corp.

Fig. 4-4. Solid-state siren circuit utilizing the LM389 IC module.

veniently implemented from such an array of active devices in a single IC package. This siren has a peak output capability of about a watt, which will suffice for some signaling requirements. For other requirements, where greater power is needed, this basic circuit can be fed into an audio amplifier with higher power capability. The undulations produced do not simulate the mechanical siren, but the psychological effect is nonetheless similar. The power of a "carrier" frequency is varied at a slow rate. The carrier can be adjusted over a range to produce the greatest psychoacoustical sensitivity. The carrier tone is intensity modulated over an adjustable range of one to seven hertz. Intensity modulation is produced by the switching action of a multivibrator on terminal 3 of the amplifier which is a tap on the

168

load resistance of the input stage. As part of this load resistance is periodically shorted out by a multivibrator-driven transistor (Q3) the self-oscillating amplifier circuit alternates its gain and power output.

The action of the multivibrator (transistors Q1 and Q2) can be inhibited by placing the single-pole switch in its off position. A sustained, high-level tone will then be produced. The tone format can be controlled by inserting a 2K variable resistance between pins 15 and 3.

SIMPLE CIRCUITS USING THE POWER MOSFET

The five circuits shown in Fig. 4-5 are applicable to many more complex systems. Each of these circuits uses the Siliconix VPM1 power MOSFET. The basic idea is that this device can be switched from substantially zero output current to 2 amperes when a positive voltage of 10 volts is applied to the gate. Compared to the base current of bipolar power transistors, the gate current of the power MOSFET can be said to be negligible. The ability to control 2

(A) Lamp driver.

(B) Relay or solenoid driver.

(C) LED driver.

(D) Low-resistance analog switch.

(E) A typical switching circuit.

Courtesy Siliconix, Inc.

Fig. 4-5. Simple circuits using the VPM 1 power MOSFET.

(A) Amplifier circuit.

Fig. 4-6. Single 40-watt channel of stereo amplifier

amperes of current from signal sources providing negligible current is a new dimension of solid-state power manipulation. Also, the freedom from secondary breakdown and thermal runaway, together with high-speed switching capability, lend support to the manufacturer's contention that these devices are destined to displace the bipolar transistor in many applications.

The circuitry for a 40-watt-per-channel stereo amplifier is shown in Fig. 4-6. The schematic diagram in Fig. 4-6A pertains to both amplifier channels. The direct paralleling technique for the two groups of three output MOSFETs is illustrated in Fig. 4-6B. (Sometimes it is expedient to place ferrite beads on the individual gate leads and/or to connect 1K resistances in series with the individual gate leads to discourage spurious oscillations.) A significant characteristic of the paralleling procedure is that no current-sharing resistances are required. This is because the power MOSFET does not attempt to hog applied current, as do bipolar transistors. Inspection of the output portion of circuit Fig. 4-6A will reveal that these devices are connected in a quasi complementary-symmetry, push-pull arrangement. The operating mode is Class AB, with just enough idling current to keep crossover distortion at an acceptable level. In the near

(B) Parallel group of output devices.

(C) Suitable power supply for operation of both output channels.

Courtesy Siliconix, Inc.

with power MOSFET output stage.

future, one can expect to have both "npn" (n-channel) and "pnp" (p-channel) power MOSFETs so that true complementary-symmetry output circuits can be used.

It is appropriate to mention that many hi-fi enthusiasts feel that the sound from a properly designed power FET amplifier is similar in quality to that previously obtained from well-engineered tube-type equipment, and is, therefore, *superior* to the sound from an amplifier employing bipolar transistors. This is a controversial subject involving much subjectivity and many variables. Quotations of performance parameters usually add more heat than light to any endeavor for objective comparison. It is true, however, that the power MOSFET design readily lends itself to the use of minimal loop feedback. Additionally, this device has an inherent frequency response into the high radio frequencies. Finally, the saturation characteristics are "soft" rather than abrupt as in the bipolar transistor. Whether all this truly leads to enhanced fidelity of sound reproduction is left to the judgement of the reader.

Other features of the amplifier schematic are conventional. The symbols labeled D1, D3, D6, and D7 are constant-current diodes fabricated with junction FETs. They substitute for the commonly used constant-current sources designed around bipolar transistors.

HIGH-EFFICIENCY LAMP DIMMER

The power MOSFET circuit of Fig. 4-7 represents an interesting departure from conventional techniques. Essentially it utilizes a vari-able-duty-cycle relaxation oscillator to control average drain current in the switched MOSFET stage. Inasmuch as the MOSFET is either on or off, dissipative losses are very low. The switching rate is ap-

Fig. 4-7. High-efficiency lamp dimmer circuit.

proximately 2 kHz. The duty cycle of the oscillator is determined by the ratio of resistance R1 and R2. Smooth control of light intensity can be obtained from manual adjustment of R2.

On the one hand, this control scheme neatly avoids the high dissipation inherent in rheostat circuits, including those in which the "rheostat" is a linearly controlled device such as a bipolar transistor. On the other hand, some of the troubles often encountered with phase-controlled thyristors will be either minimal or absent. Specifically, one is not likely to experience such shortcomings as hysteresis, lamp flicker, rfi[1], temperature effects, or performance variations with different semiconductors. The fact that operation is forthcoming directly from a 12-volt dc supply is also noteworthy. The imaginative experimenter will perceive other potential uses for such a control technique since load control is not limited to lamps. The circuit may also form the basis for other applications, such as audio alarms and switching type regulators. The enable/inhibit function of the free gate terminal enhances the versatility of the circuit. It should also be kept in mind that a number of such power MOSFETs can be directly paralleled for greater load current capability.

POWER IC LAMP FLASHERS

Although lamp flashing applications are often designed around thyristors, useful circuits can be constructed with other solid-state power devices. The lamp flashers shown in Fig. 4-8 are interesting in that a power IC is utilized to control lamp current. This IC, the LM195/LM295, is drawn schematically as a transistor. It is known as a *synthesized transistor* and has terminals corresponding to the base, emitter, and collector of an npn power transistor. However, this is the extent of the resemblance since the IC device possesses some very unique features. One of its salient characteristics is its internal self-protection against excessive current, power dissipation, or temperature rise. For example, inadequate heat-sinking would merely cause an automatic "shut-down" rather than destruction. Also the usual SOA curve pertaining to ordinary transistors is not relevant here inasmuch as internal protection guards against secondary break-

---

[1] A sensitive communications receiver in close proximity to the controller will respond to numerous harmonics of the 2-kHz repetition rate inasmuch as rise and fall times are quite fast. However, the shortcoming of *thyristor*-generated rfi is that the harmonics tend to back up into the ac power line, which then liberally distributes the high-frequency energy via radiation, conduction, and inductive coupling. Such behavior could conceivably plague the MOSFET circuit if it were operated from a power supply connected to the ac line. In general, such trouble will be much less severe than with thyristor circuits because of the "cleaner" operation of the gate oscillator compared to thyristor triggering.

(A) Flasher using a self-oscillating circuit.

(B) Lamp drive circuit controlled by CMOS or TTL logic.

Courtesy National Semiconductor Corp.

Fig. 4-8. Flasher and lamp drive circuits using a synthesized transistor.

down. Moreover, the "base" can be driven up to 40 volts without damage.

Important as is the inherent self-protection, the circuit usefulness of this power IC is mainly due to its tremendous power gain—several microamperes drive is sufficient for collector current saturation. It is not surprising that these many features are obtained, for this IC contains more than 20 equivalent transistors on a monolithic chip. What may be surprising, though, is its cost competitiveness with conventional power transistors of comparable power and current rating.

The circuit of Fig. 4-8A performs as a multivibrator, with the transistors alternating their conductive states. If both active elements were conventional transistors, one could argue that oscillation would not begin because the application of dc power would find neither transistor forward biased. The synthesized transistor, however, is in some operational respects similar to certain vacuum tubes and FETs. Specifically, when the base circuit of the LM195 is open (transistor Q1 is not turned on), full current is delivered to the lamp, although this is not revealed by the symbol of the LM195. However, there is a constant-current generator connected internally between the collector and base. The constant-current generator behaves as an extremely high resistance and enables the input impedance of the device to be very high. At the same time, the open-base operating condition is not vulnerable to static charges and is relatively immune to stray pick-up of electrical interference. In any event, when the battery voltage is connected, transistor Q2 immediately supplies lamp current and starts an exponential voltage build-up across capacitor C1. When this voltage becomes sufficient to turn transistor Q1 on, transistor Q2 is turned off and the lamp is extinguished. The current path provided by the base-emitter circuit of Q1 depletes the charge in C1, whereupon transistor Q2 and the lamp turn on again. This cycle is repetitious. As shown, the circuit produces about 45 flashes per minute. (A 500K variable resistance connected across capacitor C1 can be used for adjustment of the duty cycle.)

The lamp circuit of Fig. 4-8B is intended to be controlled by CMOS or TTL logic. Here the high base impedance of the synthesized IC transistor is a marked contrast to the power-demanding input of conventional transistors. Another useful feature of this IC is that several of them can be directly paralleled in order to drive large lamp loads.

A SIMPLE BATTERY CHARGER

Although automotive type 12-volt batteries may be charged from primitive sources of dc power under duress, certain refinements are generally desired. The battery charger shown in Fig. 4-9 has features that may not be initially apparent from the simplicity of the circuit.

Courtesy RCA Corp.

Fig. 4-9. Battery charger with automatic rate control and other protective features.

This charger is capable of sustaining a three- or four-ampere charge rate until the battery attains a substantially full charge, whereupon the charge rate is automatically reduced to a much lower level. If, during the charging process, the ac power should be switched off, or otherwise interrupted, there will be virtually no drain imposed on the battery, even though it remains connected to the charger. Also, the charger will not be damaged by either shorts or reverse-polarity battery connections. Although most or all of these features have been accomplished by various other circuits, the circuit of Fig. 4-9 is very favorable with regard to cost, ruggedness, and performance.

The basic configuration of this charger is that of a voltage-regulated power supply. Two differences from ordinary regulator operation arise from the fact that a "live" rather than an essentially resistive load is involved and that the CA3094A IC is not an ordinary operational amplifier. Specifically, terminal number 5 provides the function of current-proportional gain control. That is, the output of this IC is not only governed by the voltage difference sensed at its differential input amplifier (terminals 2 and 3), but by the current flowing into terminal 5. (Internally this current controls the supply of current to the differential amplifier and, thereby, the transconductance of its pair of transistors.) Yet another feature of this IC is that its current and power capability definitely place it in the category of power devices. It can deliver a steady current of 100mA and a peak output current of 300 mA. This also corresponds to a sustained load power of 3 watts. This IC is thus able to directly drive a medium power transistor such as the 2N3054. The internal dissipation of this device is only a few microwatts in the quiescent state.

The salient operating feature of this charger is that it samples the battery voltage and shuts down the charging current when the battery develops 14 volts across its terminals. This is accomplished via the 100K resistance connected to terminal 2, the inverting input. In this regard, the op amp behaves as a voltage comparator, with the 14-volt zener diode providing the reference voltage. On the other hand, *during* the charging process while the battery voltage is *below* 14 volts, charging current passes to the battery.

FAST CHARGERS FOR NICKEL-CADMIUM BATTERIES

The advent of the rechargeable nickel-cadmium battery has resulted in "cordless freedom" for many electrically operated products. Included are shavers, soldering irons, transceivers, hand drills, toothbrushes, food mixers, carving knives, vacuum cleaners, and clothes brushes. Although the convenience gained through the use of battery power is obvious, a major *inconvenience* has also accompanied this new mode of operation. It is commonly required that a charging process of about 15 hours be applied to restore a discharged battery to full capacity. Unfortunately, the nickel-cadmium cells are not as amenable to "quick charging" as are the lead-acid cells that are used in automobile batteries. It has, however, been determined that fast charging is practical providing that the charging current is turned off (or reduced to a "trickle-charge" rate) as soon as a fully charged condition is reached. Otherwise, there will be an abrupt rise in cell temperature together with the emission of gas. This may result in rupture of the cell and is otherwise undesirable because it tends to shorten the life span of the cell in terms of the total number of charge-discharge cycles the cell can provide. Venting is not a practical solution, inasmuch as the lost electrolyte diminishes the capacity of the cell.

Two experimental circuits that enable considerable reduction in charging time are shown here. It appears that a one-hour charging

(A) Attainment of fully charged condition in an ordinary nickel-cadmium battery.

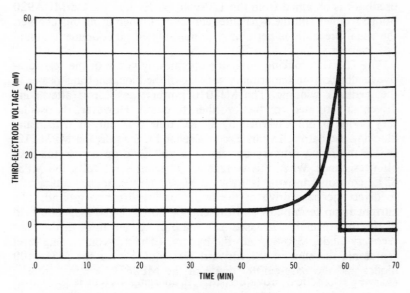

(B) Attainment of fully charged condition in a special third-electrode nickel-cadmium battery.

Courtesy Motorola Semiconductor Products, Inc.

Fig. 4-10. Cell voltage changes for ordinary and special-type nickel-cadmium batteries.

period is feasible in some cases. Under laboratory conditions, even greater acceleration of the charge time is possible. The difficulty involved is the *detection* of some cell parameter that indicates a full charge. Cell *voltage* has thus far proved more reliable than other parameters, such as temperature or impedance. Fig. 4-10 is a plot of cell voltage as a function of charge time for ordinary nickel-cadmium cells and for a special type nickel-cadmium cell provided with a *third* electrode. It can be seen that exceptionally stable and precise detection is needed for determining a full-charge condition in ordinary cells. Somewhat balancing the detection problem, it has been found that all nickel-cadmium batteries are more receptive to *pulsed* current than to straight direct current. This is fortunate due to the efficiency and controllability of electronic pulsed-energy systems.

The circuit of a 25-kHz battery charger for conventional nickel-cadmium batteries is shown in Fig. 4-11. This system consists of two functional sections, one being the 25-kHz "power supply," and the other comprising the full-charge detection circuitry. The power supply will be discussed first.

The 25-kHz pulsed current is generated by a blocking oscillator utilizing the MJ3010 power transistor, the three-winding transformer, and associated circuitry. The dc operating power for this blocking oscillator is obtained from the 120-volt, 60-Hz line via the MDA920 bridge rectifier and the 8-μF filter capacitor. (When using this charger, the usual precautions pertaining to nonisolated ac equipment should be observed.)

The 25-kHz blocking oscillator operates by virtue of the regenerative feedback from the primary winding of the blocking transformer to the feedback winding. The MJ3010 power transistor is alternately driven to saturated conduction, then to cutoff. However, its oscillatory period is not governed by an RC time constant, as with many blocking oscillators. In this circuit a second transistor, the MPS6512, senses the rising primary current as a voltage drop across the 3.0-ohm resistance. When this voltage drop becomes 1.2 volts, the MPS-6512 control transistor is forward biased and conducts sufficiently to divert the base drive of the oscillator transistor to ground. The current ramp terminates and the energy stored in the magnetic field of the transformer is released as a pulse of current through the fast-recovery diode, MR880, into the battery. (The 1.2-volt sensing level represents the sum of the forward biases of one of the MSD6100 diodes and the base-emitter diode of the MPS6512 control transistor.) The cycle then repeats, with a new current ramp produced in the primary winding of the transformer.

The presence of the MSD6100 diode pair is due to the fact that the control transistor is also used to sense the fully charged condition of the battery and it is desired to keep the two sensing functions iso-

Courtesy Motorola Semiconductor Products, Inc.

Fig. 4-11. Fast charger for conventional nickel-cadmium batteries.

179

lated from each other. In this second sensing function, the conrtol transistor inhibits oscillation as long as the terminal voltage of the sampled battery cell remains appreciably higher than it is when the cell is less than fully charged.

A transient-suppressor network is connected to the base of the oscillator transistor to protect it from destructive spikes during the turn-on portion of the cycle. This network consists of the 1N4003 diode, the 3.9-megohm resistance, and the 0.1-μF capacitor.

The sensing circuit is configured around the MC789P hex inverter. This RTL integrated circuit provides six independent inverter stages admirably suited to the requirements of this charger. The cell voltage is monitored through the 50-ohm potentiometer by inverter B. The base-emitter diode of this stage provides the threshold level as well as the temperature compensation for the sensing circuit. Amplification of the output from inverter B is provided by inverter C. Inverters D and E comprise a flip-flop. Through their action the 25-kHz oscillator is kept inoperative after the battery becomes fully charged. Inverter F is a final buffer stage which "reports" the status of the sensing circuit to the MPS6512 control transistor in the 25-kHz oscillator circuit. Note that we have not yet accounted for inverter A. Inverter A short-circuits the output of inverter C during the time interval in which the battery is accepting charging current. It accomplishes this indirectly by sampling the collector voltage of the MJ3010 oscillator transistor. Because of this circuit action, the sensing of the cell voltage can only control the charger when no current pulses are being delivered to the battery. This provides a more realistic monitoring of the charge status of the battery.

The hex inverter module receives about 2.5 volts of regulated dc operating power from the two series-connected MZ2361 stabistor diodes. Although forward conducting, these diodes regulate in a manner similar to that of zener diodes. The 0.01-μF capacitors connected to the hex inverter provide required electrical noise immunity to the sensing circuit.

When the charger is placed in operation, some initial experimentation will be necessary to determine the proper setting of the sensing adjustment potentiometer. After the correct setting has been found for a given type of battery, good performance with fast charge rates and reliable turn-off despite changes in line voltage and temperature should prevail. If, however, the special nickel-cadmium batteries with third-electrode sensing provision are available, the charger to be described next is recommended.

The fast charging circuit shown in Fig. 4-12 is intended for nickel-cadmium batteries with third electrodes. As shown in Fig. 4-10, the third electrode provides a much larger voltage *differential* for the electronic sensing of the fully charged condition. Accordingly, the charger of Fig. 4-12 employs a 25-kHz current pulsing circuit similar to that of Fig. 4-11, but it is able to provide the turn-off function with a simpler sensing circuit. At the same time, the simplicity of the sensing circuit incorporating the MD8001 monolithic transistor pair can be deceptive. Attempts to simulate this circuit with discrete

TRANSFORMER DATA
CORE
CODE: 2616P-L00-3B7
AIR GAP: 0.009 INCH
MANUFACTURER:
  FERROXCUBE CORP. OF AMERICA
  SAUGERTIES, NEW YORK

COIL
PRIMARY: 168 TURNS, AWG 29 WIRE, HF
SECONDARY: 3 TURNS, AWG 19 WIRE, HF
FEEDBACK: 84 TURNS, AWG 40 WIRE, HF

Courtesy Motorola Semiconductor Products, Inc.

Fig. 4-12. Fast charger for special nickel-cadmium batteries with third electrode.

transistors are bound to prove unsuccessful. This is because only monolithic fabrication can provide two requirements of this circuit that are *very closely matched transconductance* and *very tight thermal coupling* between the transistors. When these requisites are met, the thermal coefficient of the sensing circuit can be calculated as a function of the emitter and collector resistance. Thus, the thermal coefficient of the sensing circuit can be made to compensate that of the battery. Qualitatively, the left-hand transistor may be considered the forward-bias supply for the right-hand transistor that operates as a constant-current source. The collector voltage of the latter transistor therefore behaves as if derived from a "programmable zener diode," with the programming being imparted by the voltage sensed at the third electrode of the special nickel-cadmium battery.

As with the previously described charger, it is desirable to sample charge status of the battery *between* current charging pulses only. This is accomplished by the MPS6512 transistor that is turned on whenever the positive current pulses are delivered to the battery, thereby diverting the output of the sensing circuit. In this way, the third-electrode voltage becomes more representative of the true charge state of the battery, and the full-charge turn-off performance of the charger is improved in both accuracy and reliability.

The same precautions with regard to the nonisolated power-line connections for the previous circuit also apply for this charger. The MJ3010 power transistor in both chargers should be equipped with a suitable heat sink inasmuch as one may wish to experiment with average charging currents of up to 4.8 amperes for the commonly encountered 1.2-ampere–hour nickel-cadmium cells.

INDUCTIVE-DISCHARGE IGNITION SYSTEM

There are several ways to classify the many different types of ignition circuits that have been used for automobiles. Whether solid-state or "conventional," almost all ignition systems can be neatly described as *inductive-discharge* types or *capacitive*-discharge types. Of the two types, the latter is probably the more elaborate, and is capable of superior energy delivery to the spark plugs even at very high speeds. It is interesting, however, that the automobile manufacturers decided to deploy the inductive-discharge type as standard equipment. It may well be that the picture will change in the next several years, but it does appear that a wise decision was made. The solid-state inductive-discharge type provides a number of superior features with respect to the conventional (Kettering) ignition systems used for so many years. The even better performance of the capacitive-discharge system is more meaningful to the "hot-rodder" or the perfectionist than to the average motorist. From the viewpoint of the manufacturers, the solid-state inductive-discharge type tends to be less costly and may be more reliable for everyday motoring than capacitive-discharge ignition systems. The capacitive-discharge system generally develops higher voltage and is more demanding with regard to quality of insulation. Also, more consideration is usually required with regard to routing of high-voltage cables, selection of the ignition coil, and choice of spark plugs. Conversely, the inductive-discharge system more readily substitutes for the conventional ignition system.

The Chrysler Corporation began equipping its cars with an excellent inductive-discharge ignition system in the 1973 models. In addition to very reliable solid-state circuitry, the breaker points and "condenser" of conventional systems were replaced by a practically maintenance-free magnetic *reluctor*. The reluctor is a toothed steel wheel that rotates in close proximity to the pole tip of a permanent magnet. Associated with the magnet is a pick-up coil. The flux variations thread through the pick-up coil as a result of a tooth moving past the pole tip of the magnet. This induces an emf that in turn actuates the electronic circuit and produces a precisely timed spark for the appropriate cylinder. There are as many teeth on the reluctor as there are cylinders in the engine. The design of the high-voltage distributor remains "conventional."

It is well to point out here that the inductive-discharge classification does not pertain at all to the reluctor and pick-up coil. Indeed, inductive-discharge ignition systems have been implemented with retention of the breaker points of conventional ignition systems. Inductive-discharge ignition systems are so-called because the energy for the production of the high-voltage spark is alternately stored and "drained" from the *inductance of the ignition coil*. That is, the required energy exists in the magnetic field of the ignition coil prior to its conversion to thermal energy in the gaps of the spark plugs. By this definition, it is evident that conventional ignition systems are also inductive.

In the circuit of Fig. 4-13, power transistor Q5 is controlled in such a way that a current path is maintained through the primary of the ignition coil except when a cylinder is to fire. At such time, Q5 is abruptly turned off and the magnetic field previously supported by the primary current collapses. This induces the high voltage in the secondary circuit which, of course, includes the appropriate spark plug. This timed spark-plug firing occurs as the result of a reluctor tooth approaching and receding from an exposed pole of the permanent magnet. Specifically, as the reluctor tooth completes its position of alignment with the pole tip and begins to recede, the induced voltage in the pick-up coil changes from positive to negative polarity, thereby establishing a circuit through diodes D1 and D2. This also turns off normally on transistor Q1 which, in turn, delivers a positive pulse through capacitor C3 to the base of normally off stage Q3. Finally, driver Q4 and output stage Q5 are both turned off from their normally on states. The resultant interruption of primary current in the ignition coil induces the high voltage for spark-plug firing in the secondary winding.

The function of programmable unijunction transistor **(PUT) Q2** was not included in the sequence of events just described. It will be noted that the cascade of switching stages are not all directly coupled. Rather, ac coupling is provided between the first stage (Q1) and the second stage (Q2) by capacitor C3. It is not desirable to cascade too many stages through direct coupling because of drift and temperature problems. The ac coupling solves this problem. In

Fig. 4-13. The Chrysler-type electronic ignition system.

a pulse amplifier, it is desirable that the charge in coupling capacitors be depleted during the interval between pulses. This is accomplished by the firing of PUT Q2. Note that Q2 is not immediately ready to fire at the instant that transistor Q1 opens and provides it with anode voltage. Rather, a *time delay* is imposed because of the charge in capacitor C4. At the elapse of this time delay, Q2 fires and dissipates much of the residual charge in coupling capacitor C3. When Q1 returns to its on state, transistor Q2 is commutated back to its off state. The entire circuit is then ready to generate another burst of high voltage for the next spark plug in the firing format of the engine.

POWER IC ADJUSTABLE CURRENT REGULATOR

Although the stabilization of voltage generally comes to mind when dealing with regulated dc power supplies, there are many applications in which the regulation of current provides more meaningful measurement data, more desirable performance, or better control. Many electromagnetic devices fall into this category, including relays, motors, solenoids, deflection yokes, etc. For example, the pull-in and drop-out current of relays is conveniently determined with a programmable constant-current source. Certain dc motors can be relied upon for predictable and repeatable performance when operated from constant current. Constant voltage is less meaningful inasmuch as the basic motor torque is a *current* function and the motor current is not tightly stabilized by operating from a constant-voltage source because the resistance of the copper windings is temperature dependent. Moreover, the voltage drop across the brushes is not closely predictable and varies erratically with load conditions and with aging.

Electrochemical processes, such as battery charging and electroplating, are similarly current functions. Here, too, better measurements and more accurate control result from the use of current, rather than voltage, regulators.

From consideration of the basic physics of other devices, a number of tailor-made applications are often found for current regulators. For example, the control of thermionic emission in such equipment as electron microscopes and mass spectrometers can lead to improved performance as well as extended filament life. Again, current, rather than voltage, is the more relevant parameter in heating such filaments. Finally, some semiconductor devices such as zener diodes and some microwave diodes can best be made to perform their functions when used with a constant-current source of power.

The current regulator shown in Fig. 4-14 can be adjusted to provide 0–3 amperes of load current. It consists basically of two ad-

Courtesy National Semiconductor Corp.

justable three-terminal voltage regulators with versatile circuit and operational features. The LM150 functions as the current regulating IC in this circuit arrangement which can provide up to 3 amperes of regulated output current. The LM117 operates as an external reference which, in *conjunction* with the internal voltage reference of the LM150, enables adjustment of load current down to zero. An auxiliary negative supply is needed, but it only has to provide 10 milliamperes of current. The input voltage can be between 10 and 35 volts.

This deceptively simple schematic, consisting of two "blocks" and several resistances, should not be construed to be anything less than a high-performance system since these blocks are the equivalent of a total of more than 60 discrete transistors. Included are such features as thermal regulation, thermal overload protection, automatic current limting (at greater than the guaranteed three-ampere output current), and SOA protection of the output transistors.

Be sure to use the indicated ICs—most other three-terminal IC regulators are optimally applied where fixed output voltage or current is needed or where a limited adjustment range suffices. The LM150 and LM117 ICs require much less ADJ terminal current than do "fixed" types. This makes it easier to adjust down to zero current. It also enhances regulation and the linearity of the resistance-to-current relationship, particularly at low load currents.

A ONE-AMPERE ±15-VOLT TRACKING REGULATOR

A valuable addition to the designer's or experimenter's laboratory is a single source of regulated ±15 volts, since many ICs require this format of operating power. Regulated power supplies constructed with discrete active devices have served this purpose satisfactorily. However, it is startling to contemplate the simplicity afforded by modern ICs, such as the three-terminal voltage regulator. When these

PERFORMANCE (TYPICAL)

	(−15)	(+15)
LOAD REGULATION AT $\Delta I_L = 1A$	40 mV	2 mV
OUTPUT RIPPLE, $C_{IN} = 3000\mu F$, $1L = 1A$	100μ Vrms	100μ Vrms
TEMPERATURE STABILITY	50 mV	50 mV
OUTPUT NOISE 10Hz $\leq F \leq$ 10 kHz	150μ Vrms	150μ Vrms

*RESISTOR TOLERANCE OF R4 AND R5 DETERMINE MATCHING OF (+) AND (−) OUTPUTS

**NECESSARY ONLY IF RAW SUPPLY FILTERS CAPACITORS ARE MORE THAN 3" FROM REGULATORS

Courtesy National Semiconductor Corp.

Fig. 4-15. An 1-ampere, ± 15-volt tracking regulator.

devices are employed, cost and parts count are dramatically reduced. But best of all, the performance exceeds that generally attainable with discrete devices.

The dual-voltage tracking regulator shown in Fig. 4-15 clearly illustrates this new circuit technique. Conspicuous by their absence are several transistors and their associated passive components. Note that there is no voltage reference circuit and no protective techniques other than the diodes connected across the outputs. These circuit functions are incorporated *within* the ICs and we need not be concerned about them. Thus, the usual headaches associated with making regulated supplies stable and reliable are avoided.

It is true that the unregulated dc sources are not shown in Fig. 4-15. However, these can be quite primitive, just so the voltage and current requirements are met. For example, they can consist of a common transformer with two 15–18-volt secondaries, appropriate bridge rectifiers, and about a 3000-μF filter capacitor for each supply. As can be seen from the performance specifications, the regulation and ripple suppression of this regulator are exceedingly good.

The LM340T and the LM7915T are similar ICs except that the former is a *positive* voltage regulator and the latter is a *negative* voltage regulator. As employed in the circuit of Fig. 4-15, the LM7915T is the *master* supply, and the LM340T is the *slave* supply. The numbering of the three leads is not the same for the two ICs. (For the LM7915T, No. 1 is the ground, or "adjust" terminal; No. 2 and No. 3 are output and input terminals, respectively. For the LM340T, No. 1 is the input terminal; No. 2 is the output terminal; and No. 3 is the ground or "adjust" terminal.) Note also that the 8-volt version of the LM340T is used. The actual output of the LM340T is boosted to 15 volts, or more precisely, to the output voltage of the LM7915T master supply. The LM301 op amp performs the required polarity inversion needed to bring about this voltage boost. The op amp also enhances the dynamic performance of the positive supply.

A POWER MOSFET 200-KHZ SWITCHING REGULATOR

For the past decade, 20-kHz switching regulators have been considered ideal. In the vicinity of this frequency, one obtains a pragmatic balance of cost, performance, and packaging compactness. At lower frequencies, there is the possibility of disturbing acoustic noise. At higher frequencies, a suitable bipolar switching transistor is likely to be expensive. And other devices, such as the filter capacitors and the free-wheeling diode, have posed stumbling blocks with regard to speed and losses. However, for further improvements in the power-to-weight ratio, it has always been desirable to go beyond the present 20-to-25-kHz switching rate. Also, at higher switching rates, certain emi problems tend to be reduced as higher frequencies are more readily suppressed with practical shielding and filtering techniques.

Switching regulators can now be economically constructed to operate at higher switching rates than what has hitherto been considered feasible. The advent of power MOSFETs with power capabilities of more than ten amperes is likely to revolutionize previous notions of design boundaries. Also, Schottky diodes and vastly improved filter capacitors increase the practicability of such high-frequency regulation. The regulator circuit shown in Fig. 4-16 is capable of providing ten amperes regulated at five volts. The 710 voltage comparator has higher speed capability than most op amps, and is deployed here as a self-excited oscillator and pulse-width modulator. Oscillation occurs via the positive feedback path provided by resistance R9. The noninverting terminal (+) of the voltage comparator samples both the dc and ac components of the output and is driven to produce a voltage nulling action at its output. Such nulling is achieved by utilizing more or less of the roughly triangular "ripple" voltage. The nulling is the consequence of duty-cycle, or pulse-width, modulation.

Courtesy Siliconix, Inc.

Fig. 4-16. A 200-kHz switching regulator.

If the dc output voltage tends to drop, *more* triangular voltage is utilized; switching "on" time becomes *longer* and the dc output voltage of the regulator increases again to its set level. The converse is true when there is a tendency for output voltage to rise and *less* of the sampled triangular wave is utilized to drive the voltage comparator to saturation. *Shorter* "on" time then causes the dc output voltage to decrease to its set level. Close voltage regulation of the dc output takes place because the error-signal nulling action is referenced to the constant five-volt source provided by zener diode D2. (No doubt, the experimenter can devise a stabler reference source than the one shown. The objective of the overall design was focused on the 200-kHz switching rate and the resulting small physical size of the major components.)

Transistor Q3 and its associated components form a soft start-up circuit. When the regulator is first energized from the 28-volt dc source, capacitor C13 passes sufficient charging current to turn on transistor Q3 which is normally off. This in turn clamps the collector

189

of transistor Q2 to ground, thereby depriving switching element Q1 of enhancement bias. The charging of capacitor C3 gradually alleviates this condition and the output voltage of the regulator rises smoothly from zero to its rated output level of five volts. In this way the possibility of a momentary, but load endangering, overshoot is prevented.

Switching element Q1 is intended to be the Siliconix VN84GA 12.5-ampere power MOSFET. However, the unique characteristics of these devices enables convenient paralleling of power MOSFETs with lower individual current ratings. For example, a group of five type 2N6657 units can be mounted on a common heat sink. The performance obtained will be similar to a single VN84GA inasmuch as the 2N6657 has a 2-ampere rating. Unlike bipolar transistors, these devices do not hog current nor participate in thermal runaway. Also, it is not necessary to insert ballast resistances, as is commonly required when paralleling bipolar type transistors. Nor is it necessary to be concerned over available drive since, for practical purposes, the power MOSFET exhibits infinite input impedance when deployed in a circuit such as Fig. 4-16. Of course, packaging, cost, and production considerations favor the use of a single device rather than its paralleled equivalent.

SOLID-STATE CIRCUIT BREAKER

Fuses, thermostatic switches, and electromechanical circuit breakers all are used for the protection of electrical and electronic equipment. There are situations, however, when these protective devices are not altogether satisfactory. Their shortcomings may involve sluggish action, indeterminate trip setting, bulky physical dimensions, cost, or the inconvenience of needing replacement.

It has long been known that automatic interruption of current or power could be accomplished with solid-state devices. However, until recently it has not been easy to obtain electrical ruggedness together with respectable voltage and current capabilities. The circuit shown in Fig. 4-17 is superior to earlier versions of electronic circuit interruptors. The improved performance is largely attained through the use of the DTS-4039 triple-diffused Darlington transistor. Also, the parts count is lower than it initially appears to be since the three op amps are part of a low cost quad IC. This circuit breaker safely operates at voltages as high as 300 volts and can interrupt currents up to 8 amperes. (These are *simultaneous* ratings!) The response time is on the order of 300 milliamperes per microsecond. Such rapid interruption can be a vital factor in saving sensitive devices from the damage or destruction often caused by short circuits when other protective devices are used.

Op-amp "A" is employed as a voltage comparator. It compares the voltage monitored across the 0.1-ohm "current-sense" resistance with a voltage provided by the 250-ohm adjustment potentiometer. The output of this comparator swings from 0 to +15 volts when the sense voltage exceeds the preset voltage from the potentiometer. The 8.2-megohm resistance connected between input and output of this stage produces a small amount of positive feedback. This in turn introduces sufficient hysteresis in the response of the comparator to discourage instability at the trip point.

Fig. 4-17. Schematic diagram for the solid-state circuit breaker.

Op-amp "B" has the circuit configuration of a binary flip-flop. A positive step from the voltage comparator causes the output of this binary stage to abruptly switch from zero to +15 volts. Note the provision for a *reset* pulse. A very narrow positive pulse with an amplitude of several volts suffices to restore operation after the fault has been cleared. The reset switch can be remotely located for convenience.

Op-amp "C" is used as a buffer or driver amplifier for the DTS-2000 drive transistor. The DTS-2000 is actually a Darlington type and is employed to divert to ground the 400 mA or so of current required by the base of the "interruptor," the DTS-4039. Thus, when the output of op-amp "C" swings positive, the DTS-4039 is deprived of forward bias and therefore opens the load circuit.

In order to simplify the schematic layout of the circuit diagram, the +15-volt supply for the op amps and for bias is not shown. It can be an IC module with about 450-mA current capability.

The 50-μF capacitor in the base-bias circuit of the DTS-4039 transistor makes the turn-off process more abrupt. Its function is analogous to the "speed-up" capacitors used with digital logic gates. The 1-mH inductor prevents an instantaneous surge of high current through the DTS-4039 transistor in the event of a short circuit at the load. The IN4722 diode across this inductor protects the DTS-4039 from the voltage transient which would otherwise be generated during turn-off.

In normal operation, the DTS-4039 transistor is in saturation. This results in relatively low power dissipation (about 12 watts, maximum) and makes the heat sink requirement minimal. The Delco 7277151 heat sink is adequate for ambient temperatures up to at least 75°C. Because of the minimal thermal dissipation requirements and the single IC, the packaging of this circuit can be made reasonably compact.

A 115-VOLT, 60-HERTZ INVERTER
FOR AN AUTOMOBILE

Inverters are very useful as sources of ac power when the primary source is a battery, as is the case for motor vehicles, boats, and aircraft. Another widely used application for the inverter is as the basic energy converter in regulated power supplies. When followed by a rectifier, the inverter becomes a "dc transformer" that allows voltage and current levels available from the primary dc source to be changed to more convenient levels. There are many high-performance inverters available operating at high frequencies, achieving high efficiency, and making use of sophisticated devices. However, there remains a continuing demand for simple inverters which can be conveniently constructed from low-cost semiconductors and other components.

The simple inverter shown in Fig. 4-18 provides practical functions and is inordinately easy to construct. It delivers about 50 watts of power which is ample for the operation of electric shavers, small soldering irons, recorders, many radios, appropriately rated fluorescent desk lamps, and small portable phonographs. The heart of the unit is a 50-watt, 24-volt, center-tapped filament transformer which should be readily attainable.

192

**Fig. 4-18. A 115-volt, 60-hertz inverter for operation from an
automobile battery.**

The original germanium transistors specified for this circuit were
type 2N176. However, many "workhorse" types will perform well
in this circuit. Suggested substitutes are the Motorola types 2N1535,
2N1540, 2N1545, 2N5894, or 2N3614. Some experimentation with
the value of feedback resistors R1 and R2 will probably be required
for optimum starting and operating performance. (Silicon pnp transistors such as the Motorola MJ490 will work, but *less* efficiently than
germanium units because the collector-emitter saturation voltage is
higher in silicon transistors!)

This circuit is unusual in that no feedback windings are used.
Rather, start-up regeneration results from the cross-connected 100-
ohm resistances. After start up, oscillation is largely governed by the
saturation characteristics of the transformer core. That is, when the
core saturates because of heavy conduction in one transistor, the
transistor continues to conduct until the energy stored in the core is
used up, whereupon that transistor abruptly turns off and the alternate transistor turns on. The build-up and collapse of the magnetic
flux in the core alternately reverses direction, thereby timing the
switching cycles. (An air-core version of such a circuit would also
oscillate, but in a mode more suggestive of a multivibrator—the L/R
ratio of the windings, rather than magnetic saturation phenomena,
would then govern the oscillation frequency.)

The output filter, consisting of L1, C1, and C2, "de-spikes" the
output waveform and also modifies its shape from square to trapezoidal. Such a waveshape is less likely to produce "buzzing" in electronic equipment such as recorders, radios, or phonographs. If the

193

inverter is intended to be used only for electric shavers, as is often the case, the filter network can be eliminated. For the sake of convenience, it is suggested that the inverter be equipped with a plug for the cigarette lighter and a conventional ac outlet receptacle.

A PROTOTYPE 60-HERTZ INVERTER

The inverter shown in Fig. 4-19 is a classic in more than one sense. It is an early example of applied power electronics, but remains the basic arrangement for numerous inverters still being made by hobbyists and electronic manufacturers. The 2N278 germanium transistors were originally employed in this circuit, but more modern germanium

R1,R2-100 OHM, 5 WATT, WIREWOUND
R3,R4-1.5 OHM,5 WATT, WIREWOUND
R5-150 OHM,5 WATT,WIREWOUND
CR1,CR2-1N4004
C1-20µF,50V ELECTROLYTIC
Q1,Q2-2N278
T1-TRANSFORMER: CORE - 1 5/16"
    STACK OF 125E.I. 0.014 SILICON IRON

HEAT SINK-ALUMINUM, 80 SQ. IN., FINNED THE THERMAL RESISTANCE OF THIS SINK SHOULD BE BETTER THAN 2.1° C/WATT.

Fig. 4-19. A prototype circuit for inverters.

power transistors may be used, such as the Motorola 2N2081 or 2N1558. Despite its design vintage, this inverter has several features which are not readily excelled even with more modern semiconductor devices. The circuit is a typical example of the single-transformer saturable-core oscillator. This configuration, with its electromagnetic feedback, is superior to the more primitive circuit of Fig. 4-18. Its frequency is more nearly a function of readily manipulated factors than is the case with the cross-coupled resistor feedback technique. Higher power levels can be attained by scaling up core dimensions and wire sizes as well as transistor types and/or heat removal. The grounded collector circuit enables metal-to-metal contact between the transistor cases and the heat sinks. The use of standard silicon E-I transformer laminations is a blessing to the home experimenter who does not have a toroidal coil winder.

Inasmuch as this inverter is presented as a reference circuit for inverters with various performance parameters, the following equations are given to facilitate design:

$$f = \frac{V \times 10^8}{4B_s \times A \times N}$$

where,

f is the frequency in hertz,

V is the voltage across one-half of the primary (center-tapped) winding,

B_s is the saturated flux density of the core in lines/cm^2,

A is the cross section of the core in cm^2,

N is one-half the number of turns of the total primary winding.

Generally, the saturated flux density will be obtained from specifications pertaining to the type of core material used. Voltage V is approximately the battery voltage, but more accurately, the battery voltage minus the collector-emitter voltage of one transistor (V_{CEsat}). For germanium power transistors this will be on the order of 0.3 volt. A typical value for silicon transistors is one volt.

Note that inasmuch as V, A, B_s, and f are determined from performance requirements, tables, or necessary "guessimates," the usual procedure will involve solving for N. Accordingly, the following permutation of the fundamental equation will be found useful.

$$N = \frac{V \times 10^8}{4B_s \times A \times f}$$

The number of turns on the secondary winding is determined as follows:

$$N_s = \frac{V_s \times N}{E}$$

where,

N_s is the number of turns on secondary winding,

N is one-half the number of turns of total primary winding,

V_s is the voltage across secondary winding,

E is the voltage applied to one-half of primary winding. (This is the battery or supply voltage minus the collector-emitter voltage drop of one "on" transistor.)

The number of turns on the feedback windings is often made to develop five or six volts rms. This permits the use of current-limiting base resistances, such as R1 and R2 in Fig. 4-19, and is ample to produce reliable starting. Computation of the number of turns for the feedback winding is performed exactly as for the secondary winding.

Wire size is often selected on the basis of 750 milliamperes of current per circular mill of wire area. Standard wire tables list the cross-sectional area of the various wire gauges in circular mills. Experience has shown that such current density tends to be both safe and economical. However, it should be noted that the cited design equations neglect the effect of wire resistance on operation. Thus, because of resistance in the primary winding, the *effective* voltage acting across N turns of the primary winding will be less than E. Also, the voltage available from the secondary winding will drop as soon as current is drawn through the secondary resistance. Because of these factors, it is customary to overwind the secondary winding by 5 percent or so. These considerations were omitted from the design equations for the sake of simplicity.

Close coupling between the primary and the feedback windings is very important in order to minimize the amplitude of voltage spikes which are the consequence of leakage inductance. One approach is to wind the primary winding directly over the core insulation, and then distribute the feedback windings over the entire length of the primary winding. Bifilar winding techniques are often used to promote tight coupling. Since leakage inductance from less than unity coupling between the primary and secondary windings also contributes to generation of destructive voltage spikes, the actual construction of the transformer necessarily entails conflicts and compromises.

The experimenter can readily adapt this basic inverter for use with larger germanium transistors, such as the MP505, or with silicon types, including Darlingtons.

SELECTING THE INVERTER VOLTAGE

In most technical literature dealing with square-wave inverters, a nominal output of 115 or 120 volts is indicated when these inverters are intended as substitutes for the 60-Hz power line. It stands to reason that some discussion is in order inasmuch as we are comparing sine waves and square waves. To begin with, let us confine our investigation to those situations where we wish to deliver power to electromagnetic devices, such as motors and relays.

When the square-wave output of an inverter is applied to an electromagnetic device, the inductance is usually sufficient to act as a fairly effective filter choke. This implies ever-increasing attenuation to the progressively higher odd harmonics that are the "building blocks" of the square wave. For practical purposes, the first harmonic, or fundamental, will be left to develop the electromagnetic torque in a motor or ac relay. In determining the amplitude of square-wave voltage that will exert the same effect on a motor as the rated sine-wave voltage, we must keep the following facts in mind:

1. The fundamental frequency of a symmetrical square wave is the frequency of a sine wave with the same repetition rate as the square wave.
2. The peak amplitude of the fundamental frequency in such a square wave is $4/\pi$ times the peak amplitude of the square wave. (The peak of the sine wave fundamental is actually *greater* than that of the square wave. This fact is derived from the Fourier Theorem of harmonic analysis.)
3. In a square wave, the peak and rms values of amplitude are identical.
4. In a sine wave, peak amplitude is 1.41 times the rms value.

If we put these facts together, it turns out that the rms (or peak) value of the square wave *needed* to do the same work in a motor or relay as the rms sine-wave voltage of these devices is given by the following equation.

$$V_{sq} = \frac{1.41 \times \pi \times V_s}{4}$$

where,

V_{sq} is the *required* rms (or peak) amplitude of the square wave,
V_s is the *rated* rms voltage of the electromagnetic device.

The equation may be numerically simplified as $V_{sq} = 1.11 \times V_s$.

Example: What should be the voltage output of a square-wave inverter intended to power a motor which is rated at 117 volts from the 60-Hz line? Answer: The inverter output voltage should be 1.11 × 117, or 129.9 volts.

In actual practice, a little experimentation may be in order since the calculated equivalence between the applied square wave and the rated sine-wave power may not prove valid in all cases. What often happens is that motor *current* (which is responsible for development of electromagnetic torque) may be a reasonably good sine wave, but motor *voltage* will, at best, be a trapezoidal wave. The implication of this is that much of the harmonic energy of the original square wave remains available to produce eddy-current and hysteresis losses in the motor. Thus, an inverter-powered motor may run hotter than it would from the 60-Hz line. These harmonics may also degrade the torque capability of the motor. Often, it is necessary to put less than rated power into the motor, and to expect less than rated mechanical output from it. However, much depends upon both the inverter and the motor.

The operation of electronic equipment from the square-wave output of inverters may tend to heat up power or isolation transformers because of eddy-current and hysteresis losses. However, unless the transformers are skimpily designed, this is not likely to be a problem. The successful use of the inverter may depend upon whether the filter capacitors in the power supply of the inverter-powered equipment operate near the peak voltage of the rectified ac. If so, the use of an inverter having the *same* rms voltage as that of the nominal 60-Hz line may prove unsatisfactory because the internal dc voltage within the equipment may then be too low. Here, a 10-percent boost in inverter output voltage may overcome the problem. Actually, this problem is not as acute as one might initially predict, since the square wave is more efficient at keeping filter capacitors charged than is a rectified sine wave. There also tends to be less background "hum" in inverter-powered electronic equipment. On the other hand, much trouble is frequently encountered from "hash" and rfi.

A HIGH-POWER INVERTER

Although simple 60-Hz inverters using germanium transistors have their obviously useful application area, different design approaches are needed in order to economically produce higher output levels. It is generally necessary to utilize higher dc input voltages, to employ silicon transistors, and to operate at higher frequencies. It is true that such an inverter may no longer constitute a power source for equipment intended for operation from the 60-Hz power line. However, an even greater application area is the use of inverters with subsequent rectification so that dc-to-dc "transformation" takes place. The combination inverter and rectifier is known as a converter, or a dc-to-dc converter, and is often the heart of regulated dc power supplies. Here, we are concerned with just the inverter portion of this important building block.

Another expedient often used when high output is desired is to employ *two* transformers. The circuit remains a self-excited saturable-core oscillator, but the output transformer is designed so that it operates in its *linear* region, as do conventional transformers used in most electronic applications. A much smaller saturable transformer is connected in the input circuit of the inverter so that basically the same magnetic switching phenomenon occurs as in the single-transformer inverter. However, the efficiency of the two-transformer cir-

TRANSFORMER CHACTERISTICS

	T₁	T₂
CORE	SQ. ORTHONAL; MAGNETICS INC. No. 52035-2A	MICORSIL 150 E/U .004; 1.5 BY 1 INCHES
PRIMARY	67 TURNS No. 23 WIRE	20 TURNS No. 14 WIRE
SECONDARY	24 TURNS No. 18 WIRE	40 TURNS No. 18 WIRE

(A) Inverter circuit.

(B) Trigger circuit for initial start-up.

Courtesy RCA Corp.

Fig. 4-20. Two-transformer inverter featuring high power and high efficiency.

cuit is superior to that of the simpler arrangement because transformer losses are greatly reduced in its output circuit. (Allowing a transformer to saturate is a strict "no-no" to the power engineer who tries for efficiencies in the vicinity of 97 percent or so. If we must use core saturation, it is much better to have it occur in a small, rather than a large, transformer.) Another very important feature of the two-transformer circuit is that its production of voltage spikes is minimal. This enables the transistors to work well within their SOA ratings.

Fig. 4-20A illustrates such a two-transformer inverter. This circuit is capable of delivering 400 watts to its load and also has good tolerance for the effects of inductive or capacitive loads. A separate start-

199

ing circuit is used because the optimization of operating efficiency so vital in high-power equipment is not always compatible with other characteristics—in this case, start-up ability (especially under load).

The nominal oscillation frequency is approximately 1 kHz. By changing the small saturable transformer, T1, the operating frequency can be varied over a wide range. For example, it is probable that different winding formats for T1 can provide oscillation from, say, 400 Hz to at least several kilohertz. This is because output transformer T2 merely matches the inverter to the load requirements. Unlike single-transformer inverters, the output transformer is *not* involved in frequency determination. Surprisingly, some electronic equipment rated for 50–60-hertz operation will perform well from a 1000-kHz source. However, this remains a relatively unexplored application area.

The trigger circuit shown in Fig. 4-20B supplies a pulse of forward bias to transistor Q2 in the inverter when switch Sw1 is placed in its "start" position. The magnitude and duration of this pulse are determined by resistor R_x and capacitor C_x. The values of these components are best arrived at by experimentation. Suggested initial values for the resistor and capacitor, respectively, are several hundred ohms and several tenths of a microfarad. The trigger circuit probably can be dispensed with if too much demand is not made on the self-starting ability over a wide temperature range and with heavy or reactive loads. In any event, it is not the two-transformer circuit, per se, that imposes the need of an external starting method. It is suggested that experimentation with the biasing circuit of the inverter may optimize self-starting ability if this is desired.

MOSFET POWER RF AMPLIFIERS

The advent of the power MOSFET introduced a new device into the domain of power electronics. Just as we are becoming acclimated to the use of these devices in such applications as audio-output stages, motor-control elements, and series-pass transistors for regulators, a new area of application as radio-frequency power amplifiers has evolved. Actually, such use could have been anticipated, for the specifications of the Siliconix VMP-1 power MOSFET indicate that it can switch one ampere in about 5 nanoseconds. Other features suggesting its capability at radio frequencies are its high input and output impedance and its immunity to thermal runaway. Moreover, its operational resemblance to vacuum tubes, which have long served as Class-B and Class-C rf amplifiers is too compelling to be overlooked.

The MOSFET is ideal for rf power applications in vhf and uhf communications where a few watts into the antenna often suffice. At the lower radio frequencies, power output stages tend toward levels of hundreds and thousands of watts. Here the MOSFET makes no bid for competition in its present development stage.

Fig. 4-21A shows the circuit of a simple Class-B (linear) amplifier which can deliver about 5 watts pep into a 50-ohm load at frequencies up to several hundred megahertz. The 10-watt pep push-pull version of this basic circuit is shown in Fig. 4-21B. Greater output power should be possible, particularly for interrupted service (cw transmitters). If the frequency is not too high, these MOSFETs can be paralleled easier than ordinarily is the case with bipolar transistors. Also, much enhanced electrical ruggedness is displayed by these devices compared to bipolar rf transistors. This is because the lack of secondary breakdown, together with a *positive* thermal coefficient, renders the MOSFET almost immune to the effects of load vswr and to the effects of detuning during experimentation and adjustment.

Some experimentation is necessary with these circuits because of the drain capacitance introduced by the heat sinks. The use of physically small copper heat sinks in conjunction with beryllium oxide insulators can help keep heat-sink capacitance down. All things considered, however, the conventional practices pertinent to high-frequency layout and construction will suffice to obtain good results with MOSFETs. Several milliwatts of input power should prove adequate for Class-B operation of these amplifiers. Somewhat greater drive will be needed with Class-C operation.

25-WATT VHF TRANSMITTER FOR MARINE AND AMATEUR 2-METER BANDS

The enhanced reliability of marine communications brought about by progress in solid-state technology is truly phenomenal. Only a few years ago, marine transmitters were 3-MHz am equipment and were predominantly vacuum-tube designs. While it is not necessarily valid to criticize tubes, low frequencies, or amplitude modulation, the combination of the three in boats left something to be desired. Whereas the construction of vhf gear is somewhat of an art as well as a science, experience and good engineering practices readily overcome the undesirable characteristics of high-frequency behavior. Even more of an art was the ability to make electrically short antennas absorb power efficiently from an otherwise satisfactory 3-MHz power stage. In many instances, a good portion of the available power was dissipated in the losses of impedance-matching networks. Other problems included the fractional-ohm antenna impedances, the grounding difficulties, and rf "all over the place" except in the antenna.

The transmitter shown in Fig. 4-22 is intended to be driven from an fm exciter with a 100-milliwatt, 50-ohm output, and with a carrier frequency in the range of 144 MHz to 175 MHz. Thus, operation

(A) A Single-ended linear amplifier with an output of approximately 5 watts pep.

(B) Push-pull linear amplifier with an output of approximately 10 watts pep.

Courtesy Siliconix Inc.

Fig. 4-21. The power MOSFET used in rf amplifiers.

202

Courtesy Motorola Semiconductor Products, Inc.

Fig. 4-22. Solid-state transmitter for the 144-MHz to 175-MHz vhf range.

203

is also feasible in the 2-meter amateur band and in land mobile radio band. Included in the rf circuit is a voltage-dropping circuit configured around the MJE2020 power transistor. This circuit is essentially a series-pass regulator with a bias adjustment (R2) which enables sufficient power supply voltage to be absorbed by the transistor so that the rf output of the transmitter can be reduced to one watt or less. Such a provision is stipulated by FCC regulations. Otherwise, the transmitter circuit is quite conventional. Resonant circuits are all lumped parameter designs, and "straight through" operation is intended within the specified frequency range. It should be appreciated that an rf power transistor such as the 2N6082 was available only to the military and space programs just a few years ago—and at a respectable price. An inexpensive device capable of 25 watts output up to 175 MHz was simply nonexistent until quite recently.

When an adequate heat sink is used, this output transistor has sufficient electrical ruggedness to be self-protected against the worst conditions likely to be encountered during tune-up, experimentation, or operation. This includes high battery voltage (15.5 volts) *together* with either a shorted or open-circuited load or load impedances anywhere between these extremes. However, such abuse can be endured only for short intervals. This is reminiscent of allowable overload periods with certain vacuum tubes.

A 200-WATT AUDIO AMPLIFIER
FOR STEREO SYSTEMS

Six npn silicon power transistors are utilized in the output stage of each amplifier channel in this high-power stereo system. In the circuit shown in Fig. 4-23, it will be seen that the push-pull output stage makes use of two groups of three parallel-connected RCA 1B05 power transistors. The output stage circuit is so-called quasi complementary-symmetry configuration. Although all the output transistors are npn types, the circuit operates *as if* the lower group of paralleled 1B05 transistors were pnp types. By such simulation, true push-pull performance is obtained and direct coupling to the speaker is feasible. Such direct coupling is beneficial to low-frequency response and dispenses with the practical difficulties of high currents through a large electrolytic coupling capacitor.

The quasi complementary-symmetry circuit is particularly desirable when a rugged and reliable transistor is available in only one polarity. It has often been a problem to devise true complementary-symmetry output stages from a diverse selection of npn and pnp power transistors. With silicon transistors, the pnp type has tended to lag behind the power capabilities of npn versions. There also remains the problem of matching electrical and thermal characteristics of comple-

204

NOTES

1. D1—D8 · 1N5391; D9—D12 · 1N5393.

2. Resistors are 1/2-watt, ± 10% unless otherwise
specified; values are in ohms.

3. Capacitances are in μF unless otherwise
specified.

4. Non-inductive resistors.

5. ▼ Provide approx. 1°C/W heat sinking per
output device based on mounting with
mica washer and ZnO thermal compound
(Dow Corning No. 340, or equivalent)
with $T_A = 45$°C max.

6. Mount on heat sink, Wakefield No. 209-AB,
or equivalent. (Alternatively, this type may
be obtained with a factory-attached integral
heat sink.)

7. Attach heat sink cap (Wakefield No. 260-6SH5E,
or equivalent) on device and mount on same
heat sink with output transistor.

NOTES:

1. 90°C thermal cutout attached to heat
sink for output transistors.

2. Power transformer: Signal 120-6(Signal
Transformer Co., 1 Junius St., Brooklyn,
N.Y. 11212), or equivalent. Use 125-volt
primary tap.

Courtesy RCA Corp.

Fig. 4-23. 200-watt quasi complementary-symmetry amplifier.

mentary transistors. In any event, the quasi complementary-symmetry
circuit has enjoyed widespread use and is capable of excellent performance.

It will be seen that the immediate driver stages for the output stage
are also 1B05 transistors. These drivers operate as Class-B amplifiers.
Going back one more stage, the lower-level drivers are 1E02 and
1E03 transistors, the first being an npn type, and the second being
a pnp type. It is by virtue of these two opposite-polarity driver stages
that the output stage is made to operate as if it were an npn-pnp
combination.

Completing the cascade are two more driver stages, each operating as a Class-A amplifier. These are the differential stage consisting of the pair of RCA 1A10 transistors, and finally, the input differential stage which includes a pair of RCA 1A11 transistors. The differential input stage provides a convenient method for injecting negative feedback from the output stage. Both of these differential stages are operated from constant-current sources in the form of individual circuits involving RCA 1A09 transistors. (When a differential amplifier is operated from a constant-current source, its common-mode rejection and dc stability are enhanced.)

In a high-power audio amplifier, the biasing of the output transistors is a very important consideration. True Class-B operation is not desirable because of crossover distortion when the transistors switch conduction states. A small amount of forwarding biasing alleviates this situation, but also degrades the operating efficiency. It is desirable to maintain a small amount of forward biasing, but it should be no greater than is consistent with an acceptable reduction in crossover distortion. This does not occur without special provisions because both the leakage current and the bias requirements of transistors are temperature dependent. A biasing provision is needed which maintains the idling current of the output stage approximately constant over a wide temperature range.

This function is provided by the so-called V_{BE} multiplier stage configured around the RCA 1A18 transistor. The base-emitter voltage of this stage is the bias source for the Class-B drivers and, thus, for the output transistors. The 1A18 transistor is mounted on the same heat sink with an output transistor. This constitutes a thermal feedback arrangement and provides sufficiently good temperature tracking to keep the idling current of the output stage nearly constant. The adjustment of the idling current is provided by the 1000-ohm variable resistance in the base circuit of the 1A18 stage. It will also be observed that this V_{BE} multiplier operates from one of the constant-current sources. Without this circuit provision for such a stage, the base-emitter voltage would not be solely a function of temperature.

We have thus far accounted for all of the functions except that provided by the two transistors and associated components enclosed within the dashed lines. This is a protective circuit which ordinarily does not alter the operation of the amplifier, but it will act to limit the maximum current consumed by the output stage. The sensing of output-stage current is accomplished via the three 220-ohm resistors in the emitter circuits of one group of paralleled 1B05 power transistors and the 75-ohm resistor connected to the collectors of the other group of paralleled 1B05 power transistors. The control of output-stage current is then dependent upon the conduction of the two transistors enclosed by the dashed lines. This is so because these

Measured at a line voltage of 120 V, TA = 25°C, and a frequency of 1 kHz, unless otherwise specified.

Power:
Rated power (8-Ω load, at rated
distortion) .... 200 W
Typical power (4-Ω load) .... 300 W
Typical power (16-Ω load) .... 130 W
Total Harmonic Distortion:
Rated distortion 0.5%
IM Distortion:
10 dB below continuous power output
at 60 Hz and 7 kHz (4:1) 0.2%

IHF Power Bandwidth:
3 dB below rated continuous power at
rated distortion .... 5 Hz to 35 kHz
Sensitivity:
At continuous power output rating ..... 900 mV
Hum and Noise:
Below continuous power output:
Input shorted .... **96 dB**
Input open .... **84 dB**
With 2 kΩ resistance on 20-ft. cable
on input .... 94 dB
Input Resistance .... 18 kΩ

Fig. 4-24. Performance data for 200-watt quasi complementary-symmetry amplifier.

transistors are "active" loads for the second driver stage. When these loads increase conduction, less gain is developed in this driver stage and the drive to the subsequent stages is automatically reduced. Because of this self-regulating action, it is not possible to drive the output stage much beyond the rated output power of 200 watts. The performance data and curves pertaining to this amplifier are shown in Fig. 4-24.

100-WATT SERVO AMPLIFIER
FOR INDUSTRIAL APPLICATIONS

The direct-coupled amplifier shown in Fig. 4-25 incorporates the general design techniques used in audio and stereo equipment. However, here the emphasis is on simplicity, reliability, and electrical ruggedness. The high-voltage transistors employed in the quasi complementary-symmetry output stage enable operation directly from rectified 120-volt ac line voltage. The DTS-430 npn transistors have maximum ratings of 400 volts for V_{CEO}, 5 amperes collector current, and 150 watts of power dissipation. They are made by a triple-diffusion process and have inordinately good second-breakdown capability. Consequently, the once-valid notions that high-voltage operation necessarily entailed fragile reliability are strictly passe. Indeed, high voltage and relatively low current operation circumvent certain prob-

Courtesy Delco Electronics

Fig. 4-25. 100-watt servo amplifier designed to operate directly from rectified ac line voltage.

208

lems which all too often crop up in low-voltage, high-current operation. In order to further instill peace of mind regarding 150-volt operation of power transistors, it is well to realize that 1500-volt transistors are commonly used in tv applications.

Output transistors Q7 and Q6 are driven by Darlington-connected drivers Q4 and Q5, respectively. Because Q5 is a pnp transistor, Q6 also operates as a pnp type even though it is actually an npn—thus the nomenclature "quasi." Resistance R10 adjusts the transconductance of the Q5-Q6 pair to have the same transconductance as that of the Q4-Q7 combination. A further provision for bringing about near identical operation from both branches of the push-pull configuration is the injection of positive feedback into the base of Q5 via C3 and the network consisting of resistors R7 and R8. (The amplifier incorporates a *net* negative feedback of 40 dB, so this technique does not adversely affect stability. It merely helps attain symmetrical outputs from both branches of the amplifier.)

The output stage develops its rated 100 watts into a 75-ohm load. This, of course, is considerably higher than the 4- and 8-ohm speaker impedances commonly encountered in audio work. Output transistors Q6 and Q7 operate in Class AB with about 20-mA idling current.

This amplifier develops a high open-loop voltage gain because Class-A driver transistor Q3 "sees" a high impedance at the bases of the Darlington drivers, Q4 and Q5.

SERVO AMPLIFIER FOR TWO-PHASE INDUCTION MOTOR

For applications of servo systems to aircraft, radar, and instrumentation systems, the two-phase induction motor is often considered advantageous to the dc motor. The *motor,* itself, is preferable to a dc motor because there is no commutator. This dispenses with brush sparking and, of course, the maintenance inherent in any machine using brushes. Also, the two-phase induction motor is very efficient and results in considerable reductions in size, weight, and cost, especially for 400-Hz or higher frequency systems. There are compelling advantages in the servo amplifier as well. For example, the fact that stages of the amplifier can be ac coupled circumvents problems pertaining to drift and zero-signal setting. In the event of unusually abusive conditions or a component failure, the ac-coupled amplifier is not as likely to suffer from the overall catastrophic destruction which occurs all too commonly with direct coupling.

Hitherto, there have been some design difficulties which have often ruled out the two-phase servo system. In circuits where the quadrature phase relationship has been established by the use of capacitors con-

Courtesy Motorola Semiconductor Products, Inc.

Fig. 4-26. Servo amplifier for two-phase induction motor.

210

nected with the motor windings, these capacitors tended to be quite expensive because they have to have high capacity and high current ratings. Electrolytic types, though apparently alleviating some problems, add others such as temperature sensitivity, high dissipation factor, aging, and limited life.

The servo amplifier for two-phase induction motors shown in Fig. 4-26 enables realization of the salient features inherent with the use of the induction motor without the problems that have previously plagued actual implementations. Because of the high-voltage power transistors in the output stage, it becomes feasible to use 115-volt motors. Note that there are no capacitors associated with the motor windings. Although the circuit shown is set up for 60-Hz systems, it is only necessary to apply the proper scaling factor to the phase-shift and coupling capacitors to adapt the circuit to 400-Hz or 800-Hz systems. (Multiply the values of these capacitors by 60/400, or 0.15, to convert to 400-Hz operation.)

The Darlington-connected transistors in the complementary-symmetry output stage operate essentially as a push-pull, Class-B power amplifier. (A *slight* amount of forward bias is applied in order to reduce the "dead zone.") When powering a size 18 motor, the output stage develops a voltage gain of 20 (26 dB) and the overall voltage gain of the amplifier is 1600 (64 dB). Average power supply currents are 70 milliamperes. Transistors Q3 and Q5 require the Motorola MS-10 heat sinks (or equivalent) in order to hold transistor case temperatures to 40°C with an ambient temperature of 25°C at full power output.

The driver stage is an MC1709 op amp. This stage is arranged to operate from a single dc supply and is connected as an inverting amplifier. Note that single-ended output suffices to drive the push-pull output section through coupling capacitors C1 and C2. Thus, no transformers are needed in the amplifier circuit.

The function of the first stage is to provide the required phase shift so that the ac voltages impressed across the two motor windings will differ in phase by 90 degrees. To accomplish this, the first stage in the amplifier must produce an even greater phase shift than this in order to compensate for the *opposite* phase shifts produced by the interstage coupling capacitors. Potentiometer R1 provides an adjustment for attaining the exact quadrature phase relationship at the motor windings. Although not shown as such, the MPSA-13 is actually a monolithic Darlington transistor. Because of the equal collector and emitter resistances, this stage provides unity voltage gain. The trade-off obtained is high input impedance and exceptional stability. A unique feature of this type of "active" phase-shifting circuit is that, unlike a simple passive RC circuit, the phase shift can be varied both above and below 90 degrees with negligible variation in

Courtesy Motorola Semiconductor Products, Inc.
Fig. 4-27. Power supply for the induction-motor servo amplifier.

output voltage. A suitable power supply is shown in Fig. 4-27. No filtering is needed in the high-voltage section.

Two techniques for bringing about closed-loop positioning control of a bi-phase induction motor are shown in Fig. 4-28. Both schemes provide follow-up sensing of the motor shaft position because of the tendency for the error signal at the input of the amplifier to null itself. At null, the control winding of the motor is deprived of excitation and no torque is developed. However, when there is an angular discrepancy between the positions of input and output shafts, the control winding is energized by a quadrature voltage. The greater the discrepancy, the more voltage is applied to the control winding and the greater is the corrective torque developed in the motor. (In Fig. 4-28B the three sections of the servo amplifier are lumped together.)

REGULATING SPEED CONTROLLER
FOR PERMANENT-MAGNET MOTOR

Before the advent of modern ceramic magnetic materials, the permanent-magnet dc motor was not extensively used. Such motors were

(A) Error signal generation by means of potentiometers.

(B) Error signal generation by means by synchro devices.

Courtesy Motorola Semiconductor Products, Inc.

Fig. 4-28. Closed-loop control methods using the servo amplifier and biphase motor.

actually larger than their shunt-wound counterparts and were subject to changes in the magnetization of their pole structures. In general, their operating characteristics offered no compelling features. However, it is another story with the newer permanent-magnet motors. They are physically small, their cost is reasonable, and their operation is superior to wound-field motors for many purposes. In particular, they have a very high starting torque and their speed is a fairly

213

Courtesy Motorola Semiconductor Products, Inc.

Fig. 4-29. Regulating speed-controller circuit for permanent-magnet motor.

214

predictable function of the torque exerted by the shaft when turning a mechanical load. This motor lends itself readily to electronic speed control. The circuit of Fig. 4-29 provides at least a three-to-one range of speed and any selected speed is regulated from zero (no load) to fully rated output torque. The speed constancy exceeds that of squirrel-cage induction motors. However, by appropriate selection of circuit components and perhaps other minor modifications, the speed regulation can be even further improved.

The control and regulating features of this circuit are derived from a pulse-width modulation technique in which the pulse repetition rate is governed by optically generated pulses. The mechanism involved in producing these pulses utilizes reflected light from black-and-white segments painted on the motor armature. (Of course, a shaft-coupled cylinder or disk can also be used.) The initial conversion of pulsed light to pulsed current is made by photo transistor Q1. Each "package" of reflected light is equivalent to a pulse of forward base bias applied to a conventional transistor. The fiber optics are shown in simplified form—best results are likely to be obtained with a bundle of fibers conveying light from the lamp to the "chopper." Also, some consideration will have to be given to the effect of ambient light on the painted surface of the armature.

When considered on a functional basis, photo transistor Q1 and amplifier stage Q2 constitute a pulse-shaping circuit. The objective is to produce pulses of equal amplitude despite variations in the reflectivity of the white stripes. These variations are inevitable due to the nonuniform conditions of the optical surface. It will be noted that the effective load resistance of stage Q2 is quite high. Collector clipping readily occurs and all output pulses are thereby "sliced" to the same voltage level. Such a train of uniform-amplitude pulses enables the next section of the circuit to operate as a true tachometer, wherein the output voltage level is proportional to pulse repetition rate and, therefore, to motor speed.

The *tachometer* circuit involves transistor Q3 and associated components. In this circuit, Q3 provides polarity inversion of the pulse train. The diode-capacitor network in the collector circuit of Q3 constitutes a charge-pump circuit in which capacitor C2 develops a positive voltage proportional to the incoming pulse rate.

Field-effect transistor Q4 shifts the dc level of the tachometer to comply with the circuit needs of the next stage which is a voltage comparator. Diode D3 serves to compensate the controller against changing ambient temperature.

In the *comparator,* transistors Q5 and Q6 function as a differential amplifier and transistor Q7 is an output stage. Note that *one* input of the differential amplifier samples the tachometer voltage, whereas the *other* input samples an adjustable "reference" voltage. The magnitude and polarity of the output voltage from the comparator depends upon the difference or "error" between the sampled inputs. The generation of the error signal is of great importance in any feedback or servo system. In this controller, the error signal is used to vary the duty cycle of current pulses delivered to the motor. This is accomplished via a Schmitt-trigger circuit which functions as a pulse-width modulator.

The *Schmitt modulator* consists of transistors Q8, Q9, and Q10. Transistors Q8 and Q9 form the basis of the Schmitt-trigger circuit, and transistor Q10 operates as an output stage. The Schmitt-trigger circuit regeneratively latches into one or the other of its two conductive states, depending upon the polarity of the error signal received from the comparator. In the overall scheme, this enables the motor to *receive* current as long as its speed is too low, and to be *deprived* of current as long as its speed is too high. This is what is implied by *pulse-width modulation* in this controller circuit.

The actual motor current is "metered" out by the Darlington arrangement comprising transistors Q11 and Q12 in the output circuit.

(A) Speed vs. torque for different supply voltages at approximately 40°C.

Fig. 4-30. Performance curves for

Diode D6 provides a motor-current path when the output stage is in its nonconductive state. This current is derived from the energy *stored* in the motor inductance during "on" time. The effect of this *free-wheeling* diode is to make the actual motor current *continuous* rather than pulsed. In a real sense, such a diode, in conjunction with the motor inductance, behaves as a *filter*.

The performance of this controller is illustrated by the curves in Fig. 4-30. It merits consideration in such applications as phonographs, recorders, and chart drives.

WIDE-RANGE VARIABLE SPEED CONTROL
FOR A 2-HORSEPOWER INDUCTION MOTOR

For many years the three-phase induction motor has been the work-horse for numerous industrial processes. Such application status is well deserved as these motors are very efficient, relatively maintenance free, and are readily reversed. An additional performance feature that

(B) Speed vs. torque at three different temperatures.

Courtesy Motorola Semiconductor Products, Inc.

regulating speed-controller circuits.

has proved to be a mixed blessing is the nearly constant speed characteristics of these motors. At no load, the shaft rotation approaches the synchronous speed of the rotating magnetic field; at full load, the speed drops a nominal 5 percent or so, depending upon design. For many applications, such characteristics are extremely useful. However, for many *other* uses, the otherwise compelling features of the induction motor are to no avail because of its inability to provide other than small amounts of speed variation. Until recently, the only suitable method of obtaining extensive speed control has been by means of a motor-alternator set. Obviously, such an installation has a very high initial cost. However, designers never lost sight of the simple fact that good speed control was forthcoming from a *variable-frequency power source*.

As may already be surmised, power electronics obligingly comes to the rescue. The equipment to be described here is essentially a variable-frequency inverter with three-phase outputs for speed control of a 2-horsepower, 480-volt, three-phase induction motor. The frequency can be manually adjusted over the range of 5 Hz to somewhat beyond 60 Hz, thereby enabling a 12-to-1 speed range to be attained (approximately 145 rpm to 1750 rpm). A very practical feature of this inverter is that the voltage it delivers to the motor is directly proportional to the frequency, which is a requirement of induction motors.

Although the dc power supply is not described, it must be capable of supplying 7 amperes and its output voltage must be controllable from zero to 750 volts. A single-phase design using a variac in the power line, an appropriate transformer, and a bridge rectifier should work well. However, a three-phase design with ganged variacs would be preferable from the standpoint of dc quality. A switching-type dc power supply, or some thyristor control arrangement, also merits consideration. In any event, it is the *manual adjustment of the dc supply voltage* that varies the frequency of the inverter and, therefore, the speed of the motor.

The block diagram of the variable-frequency inverter is shown in Fig. 4-31. Some of its salient features are as follows:

1. Motor voltage is directly proportional to frequency.
2. Output transistors are operated in switched mode for optimum efficiency.
3. The three-phase voltage format for the motor is processed by CMOS digital logic circuits.
4. A six-segment waveform is synthesized which is nearly as good as a sine wave for the motor.
5. "Deadzones" are inserted in the waveforms so that overlapping conduction of output transistors cannot occur.

Courtesy Delco Electronics

Fig. 4-31. Block diagram of variable-frequency inverter for speed control of three-phase induction motor.

219

6. The inverter power section is separated by optoisolators from the logic section.
7. Current limiting circuitry protects the motor, the inverter, and the power supply from overloads.
8. Reverse rotation can be accomplished via appropriate switching at logic level.
9. ICs and Darlington transistors are advantageously used to minimize the parts count.
10. There are no commutation problems, as there are with thyristors.

Before dealing with the actual circuitry of the speed-control inverter, some basic insights can be acquired by taking a preliminary look at the waveform diagram in Fig. 4-32. The three lower waveforms represent idealized currents delivered to the stator windings of the three-phase motor. These waveforms appear to be a far cry from sinusoidal shape. A question might well arise with regard to the possibility of utilizing square waves, which are much easier to generate. However, this peculiar six-segmented wave possesses three very

Fig. 4-32. Logic timing diagram for variable-frequency speed-control inverter.

desirable features. First, the third-harmonic component, which produces much of the hysteresis and eddy-current losses when square waves are used, is almost negligible in this unique waveform. Secondly, the filtering action of the motor inductance is sufficient to make the *actual* motor current smooth enough that the motor behaves nearly as well as with sine waves. (It is the current wave, rather than the voltage wave, which manifests itself as electromagnetic torque in the motor.) Thirdly, the *deadzones* protect the output transistors from the destructive effects of overlapping conduction. Because of the three-phase format, the motor suffers very little torque depletion due to these chopped waves. Such a motor is analogous to a multicylinder automotive engine in which nearly constant torque is developed during rotation of the output shaft.

Now that we are mindful of the ultimate objectives of this system, let us inspect the circuit and see how the desired operation is brought about. The voltage-controlled oscillator and three-phase logic circuitry are shown in Fig. 4-33. (Only one such circuit is required for the system.) The voltage-controlled oscillator is the 2N2646 unijunction stage. Note that, whereas base No. 2 goes to a fixed 8-volt source, the emitter circuit receives its voltage from the 0–750-volt main supply. This enables the UJT to produce pulses with a repetition rate which increases and decreases proportionally to the applied dc voltage. In this way, a 60-Hz, 480-volt motor performs satisfactorily down to a frequency of 5 Hz and a voltage of 40 volts.

All of the gates and flip-flops are CMOS logic. They exhibit high immunity to false response from the transients and noise pulses generally encountered in industrial environments. Integrated circuit IC13 operates essentially as a buffer, or driver, stage. Flip-flop IC1 then divides the UJT rate by two and provides the clock-signal pulses on which all subsequent waves are based. This important pulse train is designated CL_1 in the logic timing diagram (Fig. 4-32) and in the logic circuitry of Fig. 4-33.

The IC9 integrated circuit is a flip-flop which performs another division by two, thereby providing pulse train CL_2, which is another building block for later fabrication of the stepped waveforms delivered to the motor. The important three-phase voltage format first appears at circuit points A, B, and C, along with the complement waveforms from circuit points A', B', and C'. All of these waveforms are the consequence of a 3-bit shift register formed by flip-flops IC10, IC11, and IC12.

Both NAND gates IC15 and IC16 are necessary to prevent the shift register from performing in a forbidden mode in which the flip-flops would toggle with every clock pulse. Finally, NAND gate IC14 produces clock pulse CL_1'. The special purpose of this clock pulse is to impose a 15-degree deadzone in the current waveforms delivered to the

Courtesy Delco Electronics

Fig. 4-33. Voltage-controlled oscillator and basic three-phase logic for speed-control inverter.

Fig. 4-34. Schematic diagram for speed-control inverter.

223

motor. As mentioned earlier, this technique ensures that no overlap of conduction can occur in the output transistors.

Referring now to the schematic diagram of the inverter, Fig. 4-34, it will be observed that the incoming signals are processed by additional digital circuitry. (This inverter is one of three identical units required to drive the motor.) Signals A and A' are applied to the D inputs of D flip-flops IC5 and IC2, respectively. These flip-flops are clocked by the CL_1' pulses. The waveform diagram reveals that such clocking produces 15-degree "deadzones" in the output waveforms. Both IC4 and IC6 provide polarity inversions needed for proper drive of the inverter preamplifier stages. The IC3 integrated circuit is associated with over-current protection to be described later.

Each inverter unit has an optoisolator, such as PC1 in Fig. 4-34. This provides electrical isolation between the positive high-voltage channel and the logic and control circuitry. Darlington stage Q2 functions as a driver for the LED within the optoisolator. Each channel has DTS2000 Darlington preamplifier stages. These stages consist of Q3 and Q4 in the top channel and Q5, Q6, and Q7 in the lower channel. The additional stage (Q5) in the lower channel closely simulates the circuit effects of the optoisolator and its driver stage, Q2, on the upper channel.

The output stages of this inverter involve pairs of DTS709 transistors. The modified Darlington connection of these output transistors enables faster and more efficient switching to be accomplished than with ordinary Darlington arrangements. Basically, the technique involves the application of negative bias to the base of the output transistors. Because of inductance in the motor, freewheeling diodes D5 and D9 are necessary, as are the snubber networks, such as R9, C1, and D4. These elements help maintain motor current and keep the operation of the output transistors within their SOA ratings.

Operating voltage for the logic circuit is provided by DVR-8, a three-terminal voltage regulator. Operating voltages for the inverter preamplifier stages are provided by the two center-tapped bridge supplies involving transformers T1 and T2. These supplies also provide the negative base bias for the output transistors.

Current-overload protection occurs through the deployment of IC3, IC7, and IC8. A NAND gate, IC8, senses the voltage drop across R18 which is a very small resistance connected in series with the negative high-voltage lead. When sufficient overload current exists, IC8 inverts the sensed voltage drop and clears flip-flop IC7. In turn, base drive to preamplifier stage Q5 is removed because of the switched state of NAND gate IC3.

Actually, the motor current interruption is not constant. Rather, it endures only momentarily until the next phase transition occurs. At that time, flip-flop IC7 is reset. If the fault has been eliminated, normal current is again delivered to the motor. Conversely, if the overload persists, the voltage drop across R18 will again trigger interruption of the motor current. This on-off cycling of the current-limiting circuit can prevail indefinitely with safe operation of the dc power supply, the inverter, and the motor. And, as soon as the overload is removed, normal motor operation automatically resumes.

This project obviously is not intended for the inexperienced hobbyist or the neophyte experimenter. However, for the advanced amateur, the veteran technician, or the electronics engineer, these circuit techniques will be found suitable for many other applications. For example, adaptations can be made for smaller motors and even single-phase induction motors where simpler and less costly power supplies will be involved. On the other hand, the power-oriented engineer will find it feasible to scale-up the design to handle motors up to 10 horsepower. This could be accomplished by paralleling output transistors and providing for an even larger power supply than is needed for the described 2-horsepower system. Delco Electronics can supply such information in application note No. 58. Such speed controls have had revolutionary implications in medium-heavy industrial processes.

WARNING

The voltage and power levels involved in this motor-control system are potentially lethal. Only those who have had relevant experience with high-voltage and high-power circuits should undertake projects of this kind.

THE HYBRID-CIRCUIT POWER MODULE

Another device making a successful bid for a respectable slice of the linear power pie is the *hybrid-circuit module*. Contrary to some prevalent notions, this circuit fabrication scheme did not become extinct with the advent of monolithic integration. Indeed, modern hybrid-circuit modules not only make use of thick- and thin-film techniques and miniature printed-circuit ceramics, but also utilize monolithic chips advantageously from both electrical and thermal considerations. The hybrid-circuit modules now on the market appear to possess superior power ratings to monolithic ICs (excluding the monolithic Darlington, which is not generally classified as an IC).

Fig. 4-35 depicts a general-purpose amplifier employing the HC2500 hybrid circuit. In essence, we have a high-power op amp with a 7-ampere peak current rating. The basic arrangement shown can be readily adapted for service in a number of applications, such as audio, ultrasonics, servo systems, solenoid drivers, driven inverters, magnetic deflection of cathode-ray tubes, regulators, and motor control. From Fig. 4-36, we see that the internal circuitry of the HC2500 is as complex as are the hi-fi and servo amplifiers discussed earlier in this chapter. Yet, the complete circuit is contained in a small hermetically sealed metal case with a volume of less than three cubic inches. Exclusive of the mounting ears, the case is approximately 2 inches square and incorporates an electrically isolated base plate for making the required thermal contact with a heavy chassis, or a heat sink.

V S	54 V
P out	60 W
Idling Current (R_B = 168 Ω	50 mA
THD	0.15%
IMD @ 50 mW	0.06%
V offset Pin 3 To Gnd.	+ 100 mV
Efficiency	64%
R L	4 Ω

Fig. 4-35. General-purpose power amplifier utilizing the HC2500 hybrid-circuit module.

The device can be operated from either a single power supply or a split power supply. Its bandwidth is specified at 30 kHz when delivering 60 watts. When operated with split 37.5-volt power supplies and adequate heat sinking, 100 watts output is feasible. This giant op amp has a high open-loop gain (approximately 70 dB), but the

*ELECTRICALLY ISOLATED FROM INTERNAL CIRCUITRY.

92CS-21077RI

Courtesy RCA Corp.

Fig. 4-36. The internal circuitry of the HC2500 hybrid-circuit module.

feedback network in Fig. 4-35 produces an actual circuit voltage gain of about 33.

As might be suspected, hybrid power modules can provide a cost-effective blend of some of the salient features of *both* discrete and monolithic devices. Because of the simplicity of implementing them into systems, they have a competitive edge over discrete devices. Also, they merit consideration at power levels where true monolithic devices are not available. Thus, hybrid power devices are likely to proliferate, rather than diminish in popularity.

5

Applications Using Thyristors

UNIVERSAL POWER CONTROLLER

In addition to versatility of application, the phase-control circuit in Fig. 5-1 has a number of other features. Because the full-wave operation of a triac is utilized, the control range extends from nearly zero power to almost the maximum power that the ac power line can deliver to the load. Full-wave control enables the use of an isolation transformer, if so desired, with freedom from the core-saturating effect of the dc component inherent in half-wave control systems. Another advantage of full-wave phase control over half-wave circuits is the improvement in the power factor of the ac line, which enhances operating efficiency. Although these facts have been appreciated ever since thyristors became commercially available for the control of power, recent improvements in triac ratings have made the implementation of full-wave phase-control feasible at power levels previously the domain of the SCR. Although full-wave SCR circuits have been used with oppositely connected SCRs in parallel arrangements, the circuitry and operation are not always as simple or as economical as would be desired.

The circuit of Fig. 5-1, unlike many low-cost light dimmers, exhibits almost negligible hysteresis. Hysteresis manifests itself as differences in the setting of the control potentiometer at which the lamp turns on and turns off. It is a nuisance, not only because of the sloppiness of control, but because a momentary drop in line voltage can cause the lamp to turn off at low light levels. If such an event occurs, turn-on can be restored only by manually advancing the control potentiometer. Hysteresis is caused by an abrupt drop in the gate-capacitor

AC INPUT VOLTAGE		120 V 60 Hz	240 V 60 Hz	240 V 50 Hz
C1		0.1 μF 200 V	0.1 μF 400 V	0.1 μF 400 V
C2		0.1 μF 100 V	0.1 μF 100 V	0.1 μF 100 V
R1		100K ½ W	200K 1 W	250K 1 W
R2		2.2K ½ W	3.3K ½ W	3.3K ½ W
R3		15K ½ W	15K ½ W	15K ½ W
PHOTOCELL CONTROL	RP	1.2K 2 W	1.2K 2 W	1.2K 2 W
SNUBBER NETWORK	CS	0.1 μF 200 V	0.1 μF 400 V	0.1 μF 400 V
	RS	100 Ω ½ W	100 Ω ½ W	100 Ω ½ W
RF1 FILTER	CF*	0.1 μF 200 V	0.1 μF 400 V	0.1 μF 400 V
	LF*	100 μH	200 μH	200 μH
RCA TRIACS		2N5567 2N5569 T4120B	2N5568 2N5570 T4120D	2N5568 2N5570 T4120D

* Typical values for lamp dimming circuits.

TERMINAL CONNECTIONS

No. 1 — Gate
No. 2 — Main terminal 1
No. 3 — Main terminal 2

Fig. 5-1. Universal triac power controller.

DIAC

Fig. 5-2. Voltage-current characteristics for the diac trigger devices used with thyristors.

voltage when triggering occurs. This effect is greatly reduced by the use of a double-time-constant control circuit involving two gate capacitors. The double-time-constant circuit in Fig. 5-1 consists of resistances R1, R2, and R3 in conjunction with capacitors C1 and C2. Also contributing to preciseness of control is the diac gate-triggering device which displays the symmetrical breakdown characteristics shown in Fig. 5-2. The use of the diac also makes triggering less temperature sensitive, and makes triacs with widely different triggering characteristics behave approximately the same in the circuit.

The turn-on of a power thyristor such as the triac is abrupt and is accompanied by harmonics and electrical noise extending well into the radio-frequency region of the spectrum. These higher frequencies tend to back up into the ac line, which can cause interference to other electronic equipment. The transfer of such interference takes place by both conduction through the power line and by radiation from it. An rfi filter, such as is included in the circuit of Fig. 5-1, greatly minimizes this phenomenon. The rfi filter is composed of a low-pass "L" section involving inductor L_F and capacitor C_F. (The filter also attenuates incoming high frequencies from the power line which, in some instances, can produce erratic triggering.) The cut-off frequency of this simple filter is in the vicinty of 100 kHz.

When inductive loads such as motors are used, the commutation of the triac still tends to occur when the current goes through zero. However, the load current and load voltage are not in phase, as in the case with resistive loads such as lamps and heaters. This results in a high rate of voltage change across the main terminals of the triac. If such a voltage change is sufficiently high, it can cause enough current to pass through internal junction capacitance to behave as a gate trigger pulse. The triac would then be turned on even though an actual gate signal is not delivered from the triggering circuit at that time. Such false triggering not only disturbs the power control function of the

230

circuit, but can be damaging to both the triac and the load. A common result of false triggering is latch-up, which is the failure to turn off at zero current. False triggering due to high rates of voltage change can occur in other ways, such as from switching at rates higher than 60 Hz, or from voltage transients originating in other devices or circuits. *Snubber* networks are often employed to slow down rates of voltage change and thereby eliminate false triggering. The snubber network used in the power control circuit of Fig. 5-1 consists of capacitor C_S and resistor R_S. This network, in conjunction with the anticipated load inductance, forms a critically damped LCR circuit. Ringing and overshoot of transient energy are largely absorbed by the resistive element of the snubber network.

Surprisingly, the power to the load can also be controlled by a photocell inserted as indicated in Fig. 5-1. The overall power amplification involved in such control is tremendous, and usually requires a number of cascaded amplifier stages in "linear" circuits. As revealed by the symbol, the photocell is of the photoconductive type, rather than a photo-voltaic type such as "solar cells." When light impinges upon the cell, its resistance decreases, thereby decreasing the peak voltage to which capacitor C2 can charge and its charging rate. Accordingly, the first effect of light will be to *reduce* the load power. When sufficient light is received by the cell, the triac will be inhibited from triggering and power to the load will be turned off. An electrical heating or lighting system could conceivably be automatically turned off at dawn by exposing the cell to light from the rising sun. Various cadmium, sulphide, cadmium selenide, and selenium cells may be used. Also, thermistors may be similarly deployed to provide response to heat. It is also possible to place these sensors in the R1 portion of the circuit so that load power is *increased* in response to light or temperature.

WARNING

The RCA isolated-stud package thyristors should be handled with care. The ceramic portion of these thyristors contains BERYLLIUM OXIDE as a major ingredient. Do not crush, grind, or abrade these portions of the thyristors because the dust resulting from such action may be hazardous if inhaled.

SIMPLE SOLID-STATE AC RELAYS

Although power transistors and thyristors can be considered as solid-state versions of electromechanical relays or contactors, the analogy tends to be more academic than practical. This is because of the lack of electrical isolation between the input and output circuits of these devices. In actual relay applications, it is comforting to know that there is considerable isolation between the control switch and the controlled circuit which may involve voltages in the kilovolts range. Such isolation is readily brought about by means of an optocoupler which usually consists of a light-emitting diode and a photosensitive device within a sealed package. When the optocoupler is used to turn on or off power transistors or thyristors, a good simulation of the erstwhile relay results. Indeed, the electronic version can be endowed with superior operating features, such as zero-voltage "contact" closure and various logic functions. Even in its simplest forms, the solid-state relay displays the advantage over physically actuated contacts in that no arcing occurs. If nothing else, this enhances reliability and longevity. Many different forms of the solid-state relay have been devised for both ac and dc control. Most of these utilize the optocoupler to achieve the desired input-output isolation.

Fig. 5-3 shows two simple solid-state ac relays. The circuit depicted in Fig. 5-3A simulates the normally open mechanical relay, whereas that of Fig. 5-3B is normally closed in the absence of input current. The use of the triac rather than an SCR for load switching is advantageous in that full-wave sinusoidal ac is delivered during the on state. (Where power considerations dictate the use of SCRs, a pair of such

(A) Normally open.

(B) Normally closed.

Fig. 5-3. Two examples of simple solid-state ac relays.

232

devices oppositely connected in parallel also yields full-wave control. However, the gate circuits then lack the simplicity made possible by the single gate of the triac.)

The 4N35 optocoupler is rated at 2500 volts rms isolation between input and output. The solid-state lamp is an infrared emitting gallium-arsenide diode. The photosensitive device is a silicon phototransistor. The H11B2 optocoupler has an isolation rating of 660 volts rms and uses a silicon Darlington phototransistor. Both devices exhibit input-output isolation resistance in the vicinity of one hundred gigaohms when the test is made with a 500-volt dc source. It is evident that safety considerations can be readily met with solid-state relays utilizing optocouplers. (Optocouplers are also available with isolation voltage ratings in the several thousands volts and higher.)

CONSTANT-BRIGHTNESS CONTROL

In photography, in various display projects, and in certain photochemical industrial processes, it is desirable to not only be able to control illumination brightness, but to maintain a constant selected brightness level. The solid-state circuit shown in Fig. 5-4 can provide such brightness stabilization at any selected brightness level throughout the

SUGGESTED TRIAC	LAMP WATTAGE MAXIMUM	
	120 V	240 V
SC136D	100 W	200 W
SC141D	400 W	800 W
SC146D	550 W	1100 W
SC151D	750 W	1500 W
SC260D	1200 W	2500 W
SC265D	2000 W	4000 W

Courtesy General Electric Co.

Fig. 5-4. Constant-brightness control circuit.

range of 100 percent to less than 10 percent of nominal lamp output. This basic circuit allows the choice of various 120-volt or 220-volt lamp sizes, since the circuit operates from *either* line voltage. The table for selecting lamps and triacs is included with the schematic diagram. Due to the very respectable amount of power that can be handled by this circuit, its range of applications can extend from indoor photography to the illumination of tennis courts and athletic fields.

In operation, light-sensing transistor L14H4 is exposed to a portion of the illumination emanating from the lamp. Such exposure constitutes optical completion of the feedback loop of the system. Therefore, this phototransistor should be physically situated in such a way that there is no interruption of or interference with the light impinging upon it, such as might be imposed by fog or by dust clouds. The conductivity of the phototransistor, which varies with the amount of light it receives, in turn varies the conductivity of the D32S4/D38S4 transistor. This transistor serves as the resistive element of the RC relaxation oscillator configured about the 2N6027 programmable unijunction transistor. The output pulse from this relaxation oscillator triggers the C103B SCR, making it conductive at varying times in the ac cycle as governed by the RC time constant of the PUT. (Recall that the D32S4/D38S4 transistor is the "R" element of the RC network.)

Before proceding further, there are two features of this circuit that merit special comment. First, the C103B SCR, though triggered on by the relaxation oscillator, is commutated to its off state twice per ac cycle. This is because the SCR constitutes the "load" for the full-wave bridge rectifier circuit made up of the four DT230B diodes. Inasmuch as there is no filter capacitor, the anode-cathode voltage of the C103B goes to zero at a rate that is *twice* the power-line frequency. The second unique feature of this circuit pertains to the PUT relaxation oscillator. It too is forced to begin its oscillatory action anew each time the rectified dc from the bridge drops to zero. The 2N5172 transistor is actually used as a *zener diode*. It not only stabilizes the unfiltered dc applied to the PUT and to the phototransistor, but also alters the waveshape in a favorable way. From what has been described, it can be seen that the actions of the PUT and the SCR are synchronized to the ac power line. Specifically, the SCR is triggered on for a *certain portion* of the alternations of the ac power line. These on times are determined by the response of the phototransistor to the light received from the illuminating lamp.

Having now traced circuit operation from the phototransistor to the C103B SCR, we will now consider how control is extended to the triac and its lamp load. This final link is straightforward enough and is a technique frequently used in power electronics. Simply stated, if the output of a bridge rectifier is shorted, or is heavily loaded, the

234

current must obviously be drawn from the ac line. In this circuit, the path of the ac line to the bridge rectifier is through the 47-ohm and the 100-ohm resistors. Consequently, when the SCR is triggered into conduction, a heavier current flows through these resistances. The resultant increase in voltage dropped across the 47-ohm resistor triggers the triac into full-wave conduction. The lamp is thereby connected to the ac line. Furthermore, the triac is phase controlled because its gate responds to the phase-controlled SCR. In this manner, the lamp current, and therefore the lamp brightness, is controlled in a closed-loop system.

Suppose that, for some reason, the light intensity begins to increase from a set brightness level. (This adjustment is provided by the 250K potentiometer.) The higher light intensity then impinging on the phototransistor *increases* its conductivity. In turn this diverts forward bias from the D32S4/D38S4 transistor, thereby *reducing* its conductivity. Thus, the PUT oscillator "sees" an *increased* RC time constant and takes a *longer* time to develop an output pulse. This means that the conductive states of both the SCR and the triac are delayed for a *longer* portion of the ac cycle and the rms value of the lamp current is *decreased,* as is its brightness. Conversely, a reduction in brightness level is counteracted by the same sequence of oppositely directed events. In this manner, any brightness level selected by the 250K potentiometer is regulated against line-voltage variations and changes in lamp characteristics.

In setting up such a control system, patient experimentation may be required to find the optimum distance between the phototransistor and the lamp. Once this distance has been determined, the brightness controller/regulator should be rigidly mounted so that the angular orientation of the phototransistor with respect to incident light cannot be inadvertently altered.

The 100-μH inductor and the RC network are included as an rfi filter to reduce injection of electrical noise from the triac into the ac power line.

SLAVE PHOTOFLASH UNIT

The heart of the photoflash circuit shown in Fig. 5-5 is a light-actuated SCR, or LASCR. This slave flash is superb for high-speed photography since the overall delay in the light burst is probably well under 100 microseconds. Several such units can be employed when photographic conditions are best served by extensive illumination. There is no need to be concerned with precise synchronization between the camera-actuated flash unit and the slave units since all slave units are triggered by the initial burst of light from the camera-actuated unit.

Courtesy Motorola Semiconductor Products, Inc.

Fig. 5-5. Photoflash slave unit using an LASCR.

The LASCR actually has both electrical- and optical-gate provisions. The electrical-gate function is very useful in controlling the optical *sensitivity* of the device. Assuming that the slave flash unit has been enabled by closing switch Sw1, energy-storage C1 charges to 300 volts within several seconds. This capacitor voltage is also impressed across the FT-106 photoflash tube. Approximately 200 volts are derived from capacitor C1 by the voltage divider consisting of R2 and R3. This voltage applied to the LASCR circuitry. The slave flash unit will remain in this state until a burst of light impinges upon the window of the LASCR. The LASCR will then be triggered in a fashion similar to ordinary SCRs when supplied with an electrical gate signal. The LASCR then "dumps" the charge accumulated in capacitor C2 across the primary winding of transformer T1. The secondary winding of the transformer develops a high-voltage pulse which triggers the photoflash tube. Although turn-off of the LASCR might occur because of the momentary discharge of capacitor C1 by the ionized photoflash tube, turn-off is ensured by the resonant action between the primary winding of transformer T1 and capacitor C2. The first reverse-bias cycle of this shock-excited oscillation commutates the LASCR. Thus, the slave unit will be ready to fire again about as quickly as the master flash has recovered.

The gate circuit can be completed either through a resistance (R4), or through a 1-henry inductor. The latter arrangement is recommended. When the resistance is used the LASCR may be vulnerable to triggering from ambient light. This can be remedied in some instances by reducing the value of resistance R4, but the sensitivity to flash actuation will also be reduced. Conversely, when the *inductor* is

236

used, ordinary ambient light "sees" the LASCR with a nearly direct short in its gate circuit and triggering sensitivity will be very low. But with the fast and intense burst of flash illumination, the inductor will behave as a high impedance, thereby imparting high triggering sensitivity to the LASCR.

SIMPLE SCR BATTERY CHARGER

The battery charger circuit shown in Fig. 5-6 bears superficial resemblance to the half-wave SCR circuits commonly used for control of lamps and heaters. However, the mode of operation is quite different from that involved in the phase-shift control of 60-Hz power. Also, the performance features are uniquely relevant to the needs of battery charging. Specifically, this charger is immune to damage from short circuits and from inadvertently reversing the battery polarity. Best of all, the charger can be adjusted to automatically shut down when the battery attains its fully charged condition.

The circuit involves a unijunction transistor oscillator that delivers audio-frequency pulses to the SCR gate while the battery is undergoing charge. This results in substantially complete half cycles of 60-Hz current being delivered to the battery. (Inasmuch as the UJT oscillator frequency is much greater than that of the power line, there is always very high probability that the triggering of the SCR will occur early in the excursion of the 60-Hz sine wave.) The SCR is self-commutated

*PRIMARY 33T #22
SECONDARY 50T #22
FEROXCUBE CORE
203 F 181 -3C3

Courtesy International Rectifier

Fig. 5-6. Simple SCR battery charger with automatic shutoff.

by the 60-Hz current; that is, conduction always ceases for 120th of a second despite the continual presence of the high-frequency trigger pulses.

As thus far described, the circuit operation is essentially that of a half-wave rectifier from the battery. Note, however, the presence of the 6.2-volt zener diode, BD. This diode cannot alter the operation of the oscillator as long as the peak-point voltage of the UJT is less than 6.2 volts. However, the peak-point voltage of a UJT increases with increasing interbase voltage. Accordingly, potentiometer R2 can be adjusted so that the required peak-point voltage will be greater than 6.2 volts. In such a case, the UJT will never be triggered and oscillation will not occur. In turn, the SCR will not be triggered and no current will be delivered to the battery from the 60-Hz power line.

In actual operation of the charger, control R2 is adjusted to enable oscillation to take place with charging current being delivered to a weak battery. As the battery accumulates charge, its terminal voltage rises. The interbase voltage of the UJT is thereby increased and its required peak-point voltage increases above 6.2 volts. With proper adjustment of control R2, the oscillation of the UJT can be made to cease when the battery charges to about 14 volts. Thus, the charger will automatically shut down when the battery is fully charged and unattended operation will not result in overcharging.

For many practical applications, transformer T2 can be a step-down transformer with a 15- to 25-volt secondary winding and a current capacity up to 15 amperes. Resistance R4 limits the maximum value of the charging current and is best determined experimentally. In most instances, it will be a low-resistance, high-wattage, wirewound type. For initial evaluation of this charger, resistor R4 can be chosen as 0.5 ohm and 50 watts. An adjustable type will, of course, facilitate experimentation.

SOLID-STATE TEMPERATURE CONTROLLER

The electronic temperature controller or "thermostat" in Fig. 5-7 is capable of greater accuracy and better operating reliability than its mechanical counterpart. The triac is switched on or off by the dc voltage level appearing at its gate. The triac turns itself off each cycle because the ac load current is directly from the power line. Thus, there are no commutation problems in such an application. The load is either fully on or it is turned off. In an actual thermal system, the SET TEMPERATURE adjustment is provided by the 50K potentiometer and the duty cycle of load power will depend upon the nature of the thermal equilibrium which can be attained in the room or system. In this respect, the operation simulates that of the bimetallic thermostatic switch.

BLACK

1.0 μF
200 V

1N4002

115 VAC
60 Hz

270
2.0 W

100 μF
25 V

WHITE

*
50K

2

3 +

LX5600/
LX5700

4 −

1

150K

5.1K

2N2905   2.0K

LOAD

RCA
40529
TRIAC

1.2K

*Set temperature
On for temperature
increase

Fig. 5-7. Temperature controller.

The operation is crisp and positive because of an internal op amp within the IC which is employed as a voltage comparator. The thermal time constant of this IC is about two or three minutes in still air. In order to minimize errors from the effect of internal heating, a small heat sink can be clipped to the package. Best results are obtained with moving air, rather than still air.

In the circuit diagram of Fig. 5-7, a 1.0-μF capacitor is used to drop the line voltage to the vicinity of 12 volts, or so. The half-wave rectifier and filter are then made up of the 1N4002 diode and the 100-μF capacitor. Such a primitive power supply is adequate because there is a shunt voltage regulator within the IC. The designated triac is a sensitive-gate type and can control up to 2.5 amperes of load current.

The performance of this simple power control circuit is much superior to that readily attainable with discrete elements in the transducer function. There are more than two dozen transistors incorporated in the LX5600 IC and the temperature-sensing technique is a sophisticated one. The basic temperature-sensing element uses the difference in emitter-base voltage of transistors operating at different current densities. The requisite close matching of this transistor pair is easily accomplished in a monolithic structure.

This circuit is amenable to the control of higher power loads. The experimenter can readily cascade driver transistors in Darlington circuits in order to provide the required gate drive for larger triacs.

This system does not deliver bursts of integral cycles of ac power to the load. Rather, the load may be switched on at random times in the line-voltage waveform. Therefore, the possibility of noise generation exists. However, unlike phase-controlled systems which generate continuous noise, a thermally actuated system such as this will only produce a short, and usually benign, burst of rfi or emi during the relatively infrequent turn-on time of the triac.

INTEGRAL-CYCLE PROPORTIONAL
TEMPERATURE CONTROLLER

At first, the temperature controller shown in Fig. 5-8 might appear to be just another version of thyristor-controlled load power with, perhaps, minor circuit modifications. The fact is, however, that the operational mode is significantly different from "run-of-the-mill" phase-

*Low temp. coefficient
**Fenwell QR51J1 100K thermistor

Courtesy Motorola Semiconductor Products, Inc.

Fig. 5-8. Proportional temperature controller using zero-switching technique.

control schemes or from simple on-off control schemes. This controller apportions whole numbers of *complete* ac cycles to the heater load by means of a zero-voltage switching technique. When a thyristor already has a gate signal applied as the ac cycle increases from its zero value, 180-degree conduction takes place. Because such triggering does not produce electrical noise, there is no need for an rfi filter in

240

this controller. Another salient feature of this control method is that it constantly yields proportionate response. This manifests itself as more accurately maintained temperature control. The over-compensatory operation yielding "too-hot" and "too-cold" temperatures, inherent in mechanical and electrical off-on systems, tends to be of negligible magnitude in proportional integral-cycle control. (No thermal hysteresis need be *deliberately* incorporated in the system.)

Because of these features of integral-cycle control, proliferation of such circuitry can be expected in the future. Already, a number of "zero-voltage switches" have made their appearance. These are sophisticated IC modules that perform the zero-switching function and often only the thyristor and a few external components are needed to complete the circuit. The circuit of Fig. 5-8 is interesting in that the zero-switching circuit is made up of *discrete* components, which provides particularly clear insight into the *nature* of integral-cycle control.

The way in which the circuit of Fig. 5-8 operates is certainly not obvious from inspection. It is convenient for analysis to initially consider SCR Q3 as just an ordinary switch. Let switch Q3 be open and let the circuit be connected to the power line. For the moment, assume that the ac voltage cycle is just crossing zero in its excursion from negative to positive polarity. A path for the charging current of capacitor C3 will then be provided by resistance R9, diodes D4 and D5, and the gate of triac Q4. Therefore, triac Q4 will trigger very close to the time that the 60-Hz voltage crosses the zero axis and will remain in conduction until self-commutated by the next zero crossing one-half of a cycle later.

So far, the circuit operation has been ordinary enough. However, the turn-on of the triac also initiates the charging of capacitor C4. This capacitor charges to the peak of the positive-going line voltage. As the line voltage declines from its peak positive value, diode D6 finds itself reverse biased by the charge stored on capacitor C4. Accordingly, capacitor C4 begins its discharge cycle through diode D5 and the gate of the triac. This gate current persists while the ac voltage crosses zero as its traverses from positive to negative polarity. Therefore, the triac is also zero-voltage triggered for the negative half of the ac voltage wave. From the sequence of events described thus far, it can be seen that the triac will continue to fire at the half-cycle zero-crossing points of the applied ac voltage wave. Substantially sinusoidal load current and voltage will exist and, insofar as concerns load power, the situation is much the same as if the load were connected directly across the ac line.

If we now consider the same initial circumstance, but with switch Q3 closed, it should be plain that the triac would *not* be triggered as the positive excursion of the ac voltage wave develops. This is because gate current is now diverted through diode D3 and switch Q3. It also should be apparent that capacitor C4 has no opportunity to charge during the positive half-cycle. Accordingly, the triac does not fire during the negative half-cycle either. We should now recognize a *unique* characteristic of this circuit—the closure of switch Q3 *inhibits* all triggering opportunities for the triac.

We will next consider what happens if, as is likely to be the case, we are unable to close switch Q3 at the onset of the rising positive voltage wave. At this point in our discussion of circuit operation, no further convenience will be served by viewing Q3 as a simple knife-switch. Accordingly, we will deal with it more realistically as an SCR capable of being turned on by a gate signal. If SCR Q3 is turned on by a gate signal at some random time during a *positive* half-cycle, the triac will remain conductive, but will turn off at the end of the voltage half-cycle. However, if a gate signal is delivered to SCR Q3 at some random time during a *negative* half-cycle, SCR Q3 will not turn on at all because it is reverse biased. We now arrive at the circuit logic whereby zero-voltage switching of the triac *must* occur whenever SCR Q3 is made to behave as an open switch. Relevantly, the gate signal to SCR Q3 can be removed at any time. The triac will be triggered in all such instances, but only when the ac voltage wave undergoes its negative-to-positive transition. To see why this is so, consider the following reasoning:

Suppose that SCR Q3 has been turned on, but that its gate signal is removed sometime during the positive half-cycle of the voltage wave. By the very nature of an SCR, Q3 will remain conductive for the entire positive half-cycle. This prevents the triac from turning on in the middle of such a positive half-cycle. Inasmuch as capacitor C4 does not have the opportunity to be charged, the triac will not turn on during the ensuing negative half-cycle either. Specifically, the triac waits until the negative-to-positive zero crossing to turn on.

Finally, suppose that the SCR gate signal is removed during the negative half-cycle of the voltage wave. The SCR is already in its non-conductive state because of reverse anode-cathode voltage. Therefore, the SCR simply *remains* as an open switch and the triac will not have its operation altered. The triac will ordinarily be on for the entire negative half-cycle and will then turn on for the forthcoming positive half-cycle. However, we can imagine yet another situation. Suppose that the ac power has been connected to the circuit during the negative portion of the voltage wave. In this instance, both the SCR and the triac will be off. (Recall that the triac cannot be on during the negative voltage excursions if it was not on during the previous *positive* excursion. This is because capacitor C4 will not have had the opportunity to charge to gate firing potential.)

The main point of these circuit responses is that power to the load can be controlled by varying the ratio of on time to off time of SCR

Q3, since this will govern "dead time" and/or the number of ac cycles delivered to the load. If the load is a device with a large thermal time constant, such as a heater, the effect of such control will be similar to that which would result from a variation of the applied voltage or current.

Having now investigated the all-important technique of *zero-voltage* switching, we are in a favorable position to use it for controlling temperature. Referring again to Fig. 5-8, a simple half-wave dc power supply is used for the control circuit. This dc power supply is made up of rectifier diode D1, voltage-dropping resistance R1, and filter capacitor C1. For the moment, circuit operation will be analyzed exclusive of the relaxation oscillator, UJT Q1. Thermistor RT, resistances R4, R5, R6, R7, and diode D2 can be seen to form the arms of a Wheatstone bridge. Transistor Q2 may be considered the null detector for this bridge network. Variable resistance R7 is set so that bridge balance prevails at a desired temperature. As room or ambient temperature increases, the resistance of thermistor RT decreases. Ultimately, transistor Q2 will be turned on and will provide gate drive to SCR Q3. This, as previously explained, turns off the triac and, therefore, the heater power. The reverse sequence of events occurs when the room temperature drops.

As thus far described, the system is workable in that the heater will be turned on and off in such a way that an average temperature will be maintained as set by the adjustment of R7. Random bursts of integral cycles will be delivered to the heater and there will be no rfi produced. However, the overshoot and undershoot of temperature can be at least several degrees, depending upon thermal conditions of the environment. Much better control resolution can be attained by *modulating* the "error" signal. This is the purpose of the UJT oscillator, Q1. It adds a sawtooth voltage to the bridge network. The period of this superimposed sawtooth signal is designed to equal 12 cycles of the power-line frequency. The sawtooth wave modulates the bridge voltage so that the net bridge voltage will be above null for a fraction of the twelve-cycle group and below null for the remaining fraction of the twelve-cycle group. The actual proportioning of on and off cycles within the twelve-cycle groups depends upon how far the temperature has strayed from nominal value. Such temperature deviation is "reported" by variations in the resistance of the thermistor. The operational tendency of such control action is for the duty cycle of the on-off periods to stabilize. Temperature is then maintained over a very narrow range.

Additional zero-voltage switching circuits are shown in Fig. 5-9. These accomplish the zero-voltage switching process with the use of specialized ICs. The proportional temperature controller of Fig. 5-9A functions similarly to the discrete-component circuit in Fig. 5-8. Here, again, the null point of the error signal is modulated by a sawtooth waveform. However, the relaxation oscillator is configured around a programmable unijunction transistor (PUT) rather than a conventional UJT device. Because of the high-wattage heater controlled by this system, the sawtooth period is relatively slow. The oven heat con-

Courtesy General Electric Co.

(A) High-power proportionate temperature controller using the General Electric PA424 IC.

Courtesy RCA Corp.

(B) Oven heat controller using an RCA zero-voltage switch IC.

Fig. 5-9. Power control circuits using

troller of Fig. 5-9B employs an RCA zero-voltage switch IC known by the acronym ZVS. Three of these zero-voltage switches are the CA3058, the CA3059, and the CA3079. The circuit shown in Fig. 5-9B does not impart modulation about the set point. Because of the closely lumped thermal "circuit" in an oven, simple zero-switching control suffices. A general-purpose zero-voltage-crossing power switch is shown in Fig. 5-9C. The operation of the MFC8070 IC is such that the *opening* of switch SW applies power to the load at the first zero-voltage crossing. If desired, an optoisolator can be substituted for the mechanical switch indicated. In this way, safe electrical isolation will be provided between the control switch and the power circuit.

Another very worthwhile feature of power control by zero-voltage switching is that the thyristor is subjected to the minimum possible rate of current change when applying power to a steady-state load. That is, di/dt is as safe as it can be. Keep in mind, however, that the systems and circuits thus far described work best with resistive loads.

FLASHERS, CHASERS, AND COUNTERS

Various patterns of blinking lights are widely used for advertising, displays, traffic lights, hazard alerts, aircraft and auto signaling, etc. Prior to the availability of reliable solid-state power devices, these functions were typically accomplished with motor-driven cams and heavy silver contacts. At their best, such mechanical systems are objectionable because of arcing and because of maintenance due to the physical deterioration of the contacts.

Of the three commonly used sequential patterns for switching lamps, the *flasher* represents the simplest technique. Here, one or two

Courtesy Motorola Semiconductor Products, Inc.
(C) On-off switch using the Motorola MFC8070 IC.

zero-voltage switching ICs.

lights alternately flash on and off at a predetermined rate. In *counters,* several lamps are arranged in a linear or other geometric array and only one lamp is kept on at a time. However, there is a continuous sequential transfer of light from one lamp to another. The process is suggestive of a "bucket brigade," although to the observer the "passed" entity appears to be light rather than buckets. The *chaser* also involves progressive movement of light. In this operational mode, the first and all following lamps *remain* on. Then, after all lamps are on, they are all simultaneously turned off and the entire process recycles, commencing with the first lamp. To the observer an ever lengthening "arrow" appears to be chasing the remaining dark segment of the display.

Flasher

The flasher circuit shown in Fig. 5-10 is representative of modern trends in power control. By utilizing dedicated IC modules, both circuit simplicity and performance are enhanced. The RCA ZVS ICs are specifically intended to accomplish zero-voltage switching with thyristors. Zero-voltage switching greatly reduces rfi and emi. Additionally, it enables the triacs to operate under relatively moderate conditions—gate signals are provided at the zero-voltage crossing of the ac cycle and the triac is spared high rates of current change. This is particularly significant in switching incandescent lamps where the cold resistance of the filament is much lower than the hot resistance. Note that no dc supply is required for these ICs inasmuch as they contain their own internal dc system.

In some respects, the zero-voltage switch may be thought of as a specialized op amp. For example, ZVS1 in Fig. 5-10 is connected to

Courtesy RCA Corp.

Fig. 5-10. Solid-state flasher with zero-voltage switching and triac lamp control.

246

an external RC network which causes it to function as an astable multivibrator. Note the connection between terminal 6 on ZVS1 and terminal 1 on ZVS2. Terminal 6 on these ICs is simply a logic-level output, whereas terminal 1 provides an "inhibit" function. Referring to the flasher circuit, the voltage at terminal 6 of IC ZVS1 is high when lamp 1 is on and ZVS2 is thereby inhibited from operation during this interval. Thus, the self-oscillating IC, ZVS1, is the *master* and ZVS2 is the *slave*. When lamp 1 is turned off at the end of one-half period of the multivibrator cycle, ZVS2 is no longer inhibited and will turn on lamp 2 at the first available zero crossing of the 60-Hz line voltage. Thus, lamps 1 and 2 alternate their on and off times.

An initial "feel" for the flashing rate may be extrapolated from the following information. If the two resistors connected in series between terminals 2 and 7 of ZVS1 are each 100K and the series RC network associated with terminals 7, 8, and 13 includes a 10K resistor and a 10-μF capacitor, the *on* time of each lamp will be in the neighborhood of 3.5 seconds.

The triacs are the RCA type T2800B which permit the use of 300-watt lamps.

Chaser

The chaser circuit of Fig. 5-11 is intended for advertising displays where the illusion of a "moving" sign is desired. This particular arrangement has the feature that the rest time during recycling of the lighting sequences is twice the duration of the intervals between the turn on of subsequent lamps as the light pattern develops. It is felt that such a rhythm is more appealing to the eye. However, this feature can be readily dispensed with if so desired.

When the system is connected to the ac power line, all of the SCRs will be initially turned off. The unijunction transistor oscillator configured about Q1 will then begin to produce the first trigger pulse. This will require an amount of time determined by the charging rate of timing capacitor C2. One might suppose that all of the SCRs will be turned on by this first trigger pulse because their gates are connected through diodes to the trigger oscillator. This, however, is not the case.

Actually, turn-on will be *inhibited* in all SCRs except SCR1. The reason for this is that a dc charge trapped in coupling capacitors C8, C9, C10, and C11 reverse-bias the gate diodes, CR5 through CR8. Indeed, it is the manipulation of these trapped charges which leads to the desired sequencing of the SCRs and, therefore, the triac lamp circuit. These trigger-inhibiting charges become trapped in the coupling capacitors under two conditions:

1. When the chaser circuit is first connected to the ac power line.
2. When the last SCR (SCR5) is turned off.

Courtesy General Electric Co.

Fig. 5-11. Chaser circuit with triac-controlled lamps.

The trapped charges in the coupling capacitors are *depleted* when a preceding SCR is turned on. Thus, each SCR involved in the sequencing pattern is "cocked" ahead of time so that its gate diode is able to conduct the trigger pulse when it comes along. The lamp-controlling triacs are, in turn, triggered by SCRs 3, 4, and 5. Triac conduction endures as long as its controlling SCR remains on. When such an SCR is turned off, its triac automatically turns off because of the commutating action of the 60-Hz line voltage.

The problem now remains of turning off the SCRs. Inasmuch as these thyristors are operated from filtered dc, they are not naturally commutating and they would tend to stay in their turned-on state indefinitely. It can be seen that SCRs 3, 4, and 5 have their anodes connected to the anode of SCR2 through a capacitor. Such interanode capacitors provide commutation to the already turned-on SCRs when SCR2 is turned on. (A turned-on SCR develops a negative-going pulse which is transferred through the interanode capacitor and momentarily deprives the already turned-on SCR of holding current, thereby turning it off.)

With the above discussion as prelude, let us now trace the circuit operation through a full cycle of lamp sequencing. When the flasher is connected to the ac power line, all thyristors are initially in their off state. When oscillator stage Q1 develops its first trigger pulse, SCR1 will be turned on. No trigger will be delivered to the other SCRs because their gate diodes (CR5 through CR8) will be reverse biased as previously explained. However, when oscillator stage Q1 produces its *second* trigger pulse, SCR3 will be turned on. Therefore, triac 1 together with lamp load No. 1 will turn on. The reason for this is that the previous turn on of SCR1 depleted the trapped charge in coupling capacitor C9, thereby disposing of the reverse bias on gate diode CR6. The turn on of SCR3 depletes the stored charge in coupling capacitor C10, thereby "cocking" SCR4 for the third trigger pulse. When the *third* trigger occurs, SCR4 is turned on, as is its triac No. 2 and lamp load No. 2. The *fourth* trigger pulse finds SCR5 receptive to a gate signal and that SCR is then turned on, as is triac No. 3 and lamp load No. 3.

Note that the turn on of lamp No. 3 produces a light pattern with all three lamps glowing. The last described event, the turn on of SCR5, depletes the reverse voltage previously stored in coupling capacitor C8, enabling gate diode CR5 to be receptive to the next trigger pulse. Accordingly, the *fifth* trigger pulse turns on SCR2, which then turns off SCR3, SCR4, and SCR5 via commutating capacitors C5, C6, and C4 respectively. At this point in the sequence, *all* lamps are off and coupling capacitors C8 through C11 have again acquired trapped charges which inhibit their gate diodes (CR5 through CR8). The flasher is now ready to undergo another cycle of operation.

It will be observed that the commencement of a new cycle of operation requires *two* trigger pulses. This accounts for the *prolonged* rest period between successive cycles of operation.

Modifications can be readily made to the mode of operation just described. For example, an additional triac and lamp load can be associated with SCR1 by connecting the triac gate through a 150-ohm resistor to the anode of SCR1. A four-lamp chaser will then be obtained with the rest period between cycles equal to the interval between turn ons. Alternately, SCR1 and its associated components can be deleted. The system will then be a three-lamp chaser with equal intervals between lamp turn ons and between pattern cycles.

Table 5-1. Recommended Triacs for Incandescent Lamp Loads

Maximum Wattage for 120 V Operation	Package		
	Plastic	Stud Mounted	Press Fit
15 Watts*	C103		
150 Watts*	C106		
700 Watts	C122	C20	C22
2500 Watts		C30	C32

*Must be used with a ballast resistor to ensure adequate surge protection.

Table 5-1 shows some triac and lamp combinations which may be used. The circuit in Fig. 5-11 is quite flexible and other than the indicated components may be employed, although it is suggested that the substituted semiconductor devices have characteristics similar or superior to those indicated. Also, any number of stages and lamps can be inserted between SCR4 and SCR5 as long as the dc supply does not become overloaded.

Counter

A somewhat similar circuit to that of the chaser is shown in Fig. 5-12. However, a few subtle differences in the circuit configuration makes this system operate as a *counter,* rather than a chaser. The lamps light up sequentially, but whenever a new lamp comes on, its predecessor turns off. The visual effect is one of a traveling light source, rather than a lengthening bar of light. If, for example, a number of lamps were arranged in a circle, one would observe a rotating light making uniformly timed angular jumps along the arc of the circle. (If such lamps were driven by a chaser circuit, the visual effect would be as if a jerky hand with a "light pen" repetitively drew and extinguished a circle of light.)

The four component sections of this circuit are readily identifiable. They are the *dc power supply,* the *display section,* the *trigger oscilla-*

250

Courtesy Motoroa Semiconductor Products, Inc.

Fig. 5-12. Counter circuit for sequentially flashed lights.

tor, and the *shift register,* or *ring counter.* The dc power supply consists of power transformer T1, the bridge rectifier involving diodes D1 through D4, and filter capacitor C1. The display section contains the indicated triacs and their associated lamps. The trigger oscillator includes junction FET Q1 and UJT Q2. The ring counter involves SCRs Q3, Q4, and Q5 and associated circuitry.

Before describing the operation of the system, mention will be made of two unusual aspects of the circuitry. It will be noted that the UJT oscillator utilizes field-effect transistor Q1 in place of the more commonly encountered resistor to charge timing capacitor C2. Actually, the Q1 circuit functions as a constant-current source, with the amount of current "metered" out to C2 determined by the adjustment of R1. (Although R1 may appear to be in the same circuit position as the conventional charging resistance, the fact that the current must flow through the FET makes a considerable difference.) Specifically, this technique enables a greater range in the trigger-pulse rate than is readily obtained with simple resistance control. The approximate range of the trigger-pulse rate for this oscillator is from one every 0.1 second to one every 8 seconds.

The second unusual feature of this circuit is the way that the SCR anodes receive their positive dc voltage. Before any of the triacs can fire, it is necessary that the SCRs have an operating dc voltage applied. This dc voltage is obtained through the gate circuit of the triacs, but in the opposite direction used to trigger the triacs. Inasmuch as a triac gate can be triggered with *either* polarity, this scheme may appear perplexing. However, the SCRs in their off states do not allow sufficient gate current to fire the triacs. Only an SCR which has *itself* been turned on can fire a triac and thereby turn on its associated lamp.

When the counter is first connected to the ac power line, the oscillator will commence operation but there will be no visual evidence of this since all the lamps will remain off. This is because the trigger pulses will not be able to pass through the SCR gate diodes, D5, D6, and D7. These diodes will be reverse biased due to their connections to the gates of the triacs. (It was previously explained how a small current passed from the positive side of the power supply through the triac gate circuits.) The SCRs are *otherwise* ready to be triggered since they have nearly nine volts across their anode-cathode terminals.

To place the circuit in operation, the PUSH-TO-START switch is momentarily depressed. This provides sufficient gate current from the discharge of capacitor C3 to trigger SCR Q3. As SCR Q3 turns on, its current demand causes the triggering of triac Q6 and lamp No. 1 is turned on. At the same time that SCR Q3 fires its triac, it also *removes* the reverse bias from gate diode D6. Thus, SCR Q4 becomes "cocked," and the *next* trigger pulse then turns on Q4. Because of commutating capacitor C5, SCR Q3 is turned off when SCR Q4 is turned on. Then SCR Q4 simultaneously fires triac Q7 and lamp No. 2 comes on as lamp No. 1 is extinguished. At the same time, the reverse bias is removed from gate diode D7. This circuit action continues with SCR Q5 and triac Q8 being turned on at the next trigger pulse. The turn on of triac Q8 is accompanied by the lighting of lamp No. 3 together with turn off of lamp No. 2. Thus, the lighting sequence has been lamp No. 1, lamp No. 2, lamp No. 3. Inasmuch as the circuitry associated with SCR Q5 connects back to that of SCR Q3, the SCR control circuit is "circular." Therefore, the lamp sequence will be continuous and will indefinitely generate the light pattern, No. 1, No. 2, No. 3, No. 1, No. 2, No. 3, No. 1, etc.

Although only three stages are shown, additional SCRs, triacs, lamps, and associated components can readily be added to the indicated circuitry. Lamps up to about 500 watts can be controlled by the triac types shown. This power level is somewhat below the rated capability of these triacs, but allowance must be made for the high initial currents required when cold incandescent lamps are turned on.

FLUORESCENT LAMP DIMMER

The control of light output from fluorescent lamps is desirable for many applications. The fluorescent lamp dimmer about to be described does not use the most sophisticated electronics available. However, the circuit is basically satisfactory and is readily amenable to modifications once the underlying principles are understood.

The block diagram illustrating the basic concept of a fluorescent lamp dimmer is shown in Fig. 5-13A. The unique characteristics of the fluorescent lamp require that its light output be controlled by *current* variation rather than by changing the lamp voltage. This is because gaseous discharge lamps establish a nearly constant ionization potential across their terminals. It would not be practical to attempt to forcibly change this inherent feature. Fortunately, current variation will naturally occur if the traditional series ballast impedance used with fluorescent lamps is retained. This is because the ballast, a high-impedance inductor, absorbs across its terminals the *difference* between the applied peak voltage and the voltage needed to maintain ionization of the lamp. It is also fortunate that SCR phase control is essentially a duty-cycle modulation technique. (It is not perfect in this respect because the peak amplitude of the "chopped" line-voltage wave is not constant. However, the lamp will remain illuminated as long as the peak amplitude is sufficient to maintain ionization. In practice, a wide range of light-intensity control is attainable.)

It is phase control of line voltage in conjunction with the ballast impedance that enables the average value of lamp *current* to be varied. It also readily supplies the *higher* voltage demanded by the lamp for starting. However, there is yet *another* requirement for such a control circuit. If half-wave phase control were used, the dc component in the controlled pulses would saturate the iron core of the ballast transformer. This would decrease its impedance and would interfere with the objective of controlling the lamp current. Therefore, what is needed is full-wave phase control, which, because of its waveform symmetry, has no dc component. Fig. 5-13B shows how full-wave control is obtained from a single SCR when it is used in conjunction with a bridge rectifier. (Of course, the experimenter may wish to use a triac instead of an SCR.)

The control circuit shown in Fig. 5-13C is essentially that of a unijunction transistor used as trigger control for an SCR. As in many other UJT-SCR applications, phase control of the SCR is achieved as the result of the time required to charge the emitter capacitor to the firing voltage of the UJT. In this circuit, the 0.2-μF emitter capacitor is charged through the 150K resistance and the 750K control potentiometer.

The 33K loading resistance, R_{LB}, is needed to promote control stability. Its actual value should probably be investigated empirically.

(A) Block diagram.

(B) Circuit for obtaining full-wave control from an SCR.

Fig. 5-13. Block diagram and circuit

(C) Complete dimmer circuit.

for fluorescent lamp dimmer

255

Unless R_{LB} is low enough to make the load appear sufficiently resistive, the SCR "sees" an inductive load and may encounter commutation problems.

The number of such lamp-ballast combinations that can be controlled by this system depends, of course, upon the current capability of the SCR and the bridge rectifier. Also, it may be found that better results are obtained from lamps that have been "seasoned" somewhat, rather than from brand-new lamps. Lamps in depressed-temperature environments, or those exposed to chilling drafts, may display unreliable ionization characteristics. They may either refuse to start or they may not provide a satisfactory range of light control. Special low-temperature lamps should be used in such instances.

A HIGH-FREQUENCY FLUORESCENT LIGHTING SYSTEM

The light-producing efficiency of fluorescent lamps is better at frequencies sufficiently high that the gas does not have time to deionize between the ac voltage alternations. More important, it is contended that such light constitutes higher quality illumination. The threshold of flicker perception from lamps operating from the 60-Hz line may be close enough to the response of the eye to cause the headaches and discomfort some people apparently experience. Also, high-frquency fluorescent lighting may prove better for certain photographic purposes.

Fig. 5-14 shows an 8-kHz inverter with sufficient power capability to handle six 80-watt instant start fluorescent lamps. This inverter uses high-speed SCRs and is intended to be driven from a variable-duty-cycle trigger source. The trigger source is shown in Fig. 5-15. Such pulse-width control is an ideal way of controlling the average current and, therefore, the light output of fluorescent lamps. The inverter and its trigger source are intended for operation from a nominal 150-volt dc source with about 4-ampere capability. Such a dc supply voltage is readily obtained by rectification of the power-line voltage as shown in Fig. 5-14B. However, this simple scheme introduces the possibility of grounding problems unless precautions are taken to "float" the lamp, inverter, and trigger circuits. A safer supply employs an isolation transformer, as shown in Fig. 5-14C. Such a transformer should have a 600 VA rating in order to provide the current demand.

The inverter is the parallel type and its operation is derived from the use of the high-speed SCRs together with a reliable commutation technique. The commutating capacitor is C3, but L1, L2, C4, C5, R1, R2, D3, and D4 are also involved in bringing about precise turn-off of one SCR when the alternate SCR turns on. Diodes D1 and D2 prevent voltage overshoots during switching transitions. Inductor L3 serves as a ballast to limit current peaks which tend to occur during

T—CORE, 8 PIECES OF INDIANA GENERAL NO. CF-602 MATERIAL 05, OR EQUIVALENT.
CROSS SECTION, 8 cm²
N1, N6 - 30 TURNS OF NO.18 MAGNET WIRE
N2, N5 - 13 TURNS OF NO.18 MAGNET WIRE, 2 STRANDS
N3, N4 - 52 TURNS OF NO.28 MAGNET WIRE, 2 STRANDS

L3 — 131 TURNS OF NO. 15 MAGNET WIRE ON ARNOLD
ENGINEERING CORE NO. A4-04117, OR EQUVIALENT.

Courtesy RCA Corp.

(A) Inverter circuit.

(B) Simple line-operated dc
power supply.

(C) Transformer-isolated dc
power supply.

**Fig. 5-14. High-frequency inverter for fluorescent lamps and two optional
power supplies.**

switching. These circuit features are more or less common to SCR inverters, but in this application they have been optimized to ensure reliable commutation with 8-kHz variable-duty-cycle operation.

An interesting aspect of this system is the use of capacitors, rather than inductors as ballasts for limiting lamp current. These capacitors have another function as well. They remove the dc component from

257

Courtesy RCA Corp.

Fig. 5-15. Trigger circuit for high-frequency fluorescent lamp inverter.

the waveform delivered to the lamps. This is why control of lamp current and, therefore, light intensity, is feasible. As the duty cycle of the inverter output wave departs more and more from 50 percent (by adjusting R5 or R11 in the trigger circuit in Fig. 5-15), an increasingly greater amount of energy is invested in the dc component. (The symmetrical square wave with 50 percent duty cycle has no dc component to begin with.) Because ballast capacitors C1 and C2 displace the zero axis in such a way as to make the dc component zero, the average current in the lamps varies with duty cycle. If it were not for this action, that is, if resistors rather than capacitors were used as the ballast elements, duty-cycle variation would have little effect on lamp operation because the net energy in the waveform delivered to the lamps by autotransformer T would remain constant. It is also evident that the use of inductive ballasts would not be appropriate in this circuit. (Inductors could be used, however, if the *control* of light intensity is not required.)

The trigger circuit of Fig. 5-15 is designed around an 8-kHz multivibrator in which transistors Q2 and Q3 are the active devices. Transistors Q1 and Q4, together with the neon lamp and associated components, comprise a voltage-regulated power supply for the multivibrator. Transistor Q4 is the error amplifier/comparator, transistor Q1 is the series-pass transistor, and the neon lamp is the voltage reference. Zener diode D4, filter capacitor C5, and resistors R8, R9, R13, R14, and R16 form a 20-volt partially regulated dc supply. Therefore, the voltage and power rating of transistor Q1 can be reasonably low.

The salient feature of the multivibrator circuit is its duty-cycle control, which is provided by potentiometer R11. Potentiometer R5 is used to adjust the duty-cycle *range*. Once this range has been determined, R5 can be replaced with a fixed resistor, if desired. The two multivibrator outputs are obtained from the secondary windings of pulse transformers T1 and T2.

COLOR LIGHT ORGAN CIRCUIT

Colored lights flickering in response to music can add another dimension to perception. Sophisticated color organs have been developed in which there are certain unique correlations between audio frequency, loudness, and the color and intensity of responsive light sources. It is even conceivable that the right kind of luminous display could create the feeling of the music with no sound impinging on the ears. However, the novelty aspect of dancing lights synchronized to music is rewarding in itself. Power electronics is very much involved because the light displays, whether they be filamentary, gaseous, or solid-state lamps, must be actuated and controlled by power devices.

T1 1:1 Pulse transformer
T2 120 V:240 V CT, 30 mA

Courtesy General Electric Co.

Fig. 5-16. Circuit for color light organ or sound-activated lights.

(Thus far, LEDs have not been used much for this purpose because the eye demands strong sources of light.)

The circuit shown in Fig. 5-16 is a basic building block for experimentation with simultaneous aural and visual effects of sound. To get the "feel" of this phenomenon, it is suggested that initial investigation be confined to the simplest situation in which a single light source is controlled by monophonic sound. Later, when one has developed insights into relationships between sound and light, more complex multicolored systems can be worked out for stereo reproduction.

The ideas incorporated in the circuit of Fig. 5-16 are straightforward and are representative of much-used techniques for lamp dimmers, motor controls, etc. A selected portion of the audio spectrum enters the circuit through the input filter network. Actually, only positive portions of audio signal enter the circuit due to the diode inserted between the filter and the emitter-follower input stage. (Although 2N2711 transistors are shown in this schematic, many npn silicon audio transistors can be used.) The unidirectional pulses from the emitter follower are then applied to the base of the second transistor which operates as an "electronic resistance." That is, its effective collector-emitter resistance is modulated by the amplitude of the applied base signal. Although it can be said that many transistors operate in this manner, it is particularly significant in this circuit because it is

260

desired to vary the rate of charge of a capacitor in a relaxation oscillator. In other words, an RC time constant is varied in accordance with audio information.

The time constant of this RC network determines how soon the triac will be triggered in the 60-Hz cycle. Thus, the audio signal is able to phase-control the triac and thereby vary the effective value of the lamp current. It should not be supposed that this occurs for each cycle of the audio signal. The 0.1-μF capacitor in the oscillator circuit cannot change its voltage instantly even though the time constant is changing. Rather, the average effect of the music over a fraction of a second or more causes the light variations. This effect is also accomplished by the thermal lag of the lamp filament. (In systems using gaseous lamps, the effect is noticeably different because of the relatively fast response of these lamps.)

The usual approach to a color-organ display is to have three or more different colored lamps. Each of these lamps is driven by a separate circuit such as that of Fig. 5-16. The filter response for each circuit is different, however. For example, a red lamp might be driven from a circuit employing a 40- to 400-Hz bandpass filter, a green lamp driven from a second circuit using a 400- to 4000-Hz filter, and a blue lamp driven from a circuit with a 4000- to 20,000-Hz filter. (The low-frequency filter can, alternatively, be a low-pass type with a 400-Hz cutoff, and the high-frequency filter can be a 4000-Hz high-pass type.) Because filters which are both economical and practical may not have steep attenuation slopes, the design is often made for much narrower bandwidths than those just suggested. This partially takes into account the fact that high-amplitude audio signals will ride through the stopband regions of simple filters. Thus, with narrower bandwidths, there will be *less* simultaneous illumination of the lamps.

A few words are in order with regard to the input filter. From the viewpoint of the experimenter, LC filters impose practical difficulties in altering response characteristics. Active electronic filters alleviate this problem. With recently available ICs containing four operational amplifiers on a single monolithic chip, the circuit complexity of active filters has been greatly reduced. Thus, it turns out that the LM124 quad op amp is ideal for electronic filters. The circuit for such a filter is shown in Fig. 5-17. The LM124 connection diagram is indicated in Fig. 5-18.

With the component values indicated in Fig. 5-17, a bandpass response will be obtained with center frequency at 1000 Hz, and a bandwidth of about 20 Hz. Once the filter is operative in this manner, it is easy to modify its behavior. In order to control bandwidth, a 500K variable resistance should be connected from point "A" to ground. Center frequency control can be attained by replacing R5 and R6 with 2-megohm controls. Preferably, these controls should be ganged

Courtesy National Semiconductor Corp.

Fig. 5-17. Basic active filter for use with color organ circuit in Fig. 5-16.

Courtesy National Semiconductor Corp.

Fig. 5-18. Connection diagram for the LM124 quad operational amplifier.

on a single shaft. Additionally, the center frequency can be shifted by changing the values of capacitors C1 and C2. However, these capacitors should be maintained equal in value.

It will probably be desirable to associate either an input or output potentiometer with the active filter in order to provide amplitude control. This is because the overall gain of the circuit changes with the adjustment of bandwidth. (Much higher gains are developed at narrow bandwidths than for wide bandwidths.) An alternative approach would be to use the fourth op amp as a variable-gain output stage.

When experimenting with the color-organ circuit, always ascertain that the filter is performing as intended before evaluating the performance of the entire circuit. Also bear in mind that electronic filters are vulnerable to overload.

CONTROL CIRCUIT FOR SHIFTING BETWEEN TWO LIGHT SOURCES

The circuit of Fig. 5-19 provides the useful feature that light output can be smoothly shifted from one lamp load to another. Known also as a *tandem dimmer,* its main application is for slide projectors. How-

NOTE: Total light level (sum of lamps 1 & 2) constant within 15%.

Courtesy General Electric Co.

Fig. 5-19. Control circuit for smoothly shifting from one source to another.

ever, the circuit can be readily used for stage effects and for advertising displays. It comprises two phase-controlled triac circuits which share the phase-adjust control, R3. The adjustment of control R3 is such that the conduction angle of one triac is advanced while the conduction angle of the other triac is retarded. Thus, lamp current can be continuously transferred from one lamp load to the other as control R3 is varied.

This lamp control scheme, like other thyristor phase-control circuits, must be able to survive the heavy current inrush caused by the low resistance of cold lamp filaments. For this reason, it is not sufficient to select triacs merely on the basis of the hot filament-current demand. Moreover, the available short-circuit current from the ac source enters into the picture. A different situation is presented by ordinary residential installations than by industrial or commercial power systems. Table 5-2 provides useful guidance in matching the

Table 5-2. Triac Selection Guide for Phase-Controlled Lamps

Device	House Wiring	Industrial/Commercial Wiring
SC35/36	360 W	180 W
SC40/41	600 W	250 W
SC141	600 W	480 W
SC45/46	1000 W	450 W
SC146	1000 W	600 W
SC50/51	1200 W	600 W
SC60	2000 W	1300 W

Courtesy General Electric Co.

triacs to various lamp loads. Note that the higher resistance of house wiring enables triacs with lower power ratings to be employed.

It is appropriate here to mention another aspect of thyristor operation in this and in other lamp-control circuits. It may appear that fusing is unnecessary because of the low probability of a shorted load. However, the very ordinary advent of a burned-out lamp can be nearly as destructive to the thyristors as a "crowbar" short circuit. This is because the parting filament may be momentarily followed by an arc.

Table 5-3. Fuse Selection Guide for Phase-Controlled Lamp Circuits

Device	I^2t (amp^2 sec)
SC35/36	7.5
SC40/41	21.0
SC45/46	53.5
SC50/51	83.5
SC60/61	565.0
SC141	83.5
SC146	83.5
C106	0.5
C45/46	2000.0
C50/52	4000.0

Courtesy General Electric Co.

Although such arcing will only endure for the remainder of the half-cycle of energization, its inception can destroy thyristors. Table 5-3 is a guide for selecting fuses which will protect certain thyristors against such destruction. The basic idea is to select a fuse with an I^2t (ampere^2 second) rating below that of the device I^2t rating. The devices with the SC designations are General Electric triacs. Those with the C designations are General Electric SCRs.

CAPACITOR-DISCHARGE IGNITION SYSTEM

The capacitor-discharge system in Fig. 5-20 commands interest for at least three reasons. First, the operational mode is classical; that is, it is not likely to evolve into something drastically different. Secondly, it allows retention of the ignition breaker points, standard on all cars prior to the advent of factory-installed electronic ignition systems. Thirdly, it provides the driver the *option* of selecting either the conventional ignition system or this electronic ignition system. This is psychologically desirable, since most of us tend to feel more secure with such an emergency provision.

It should be pointed out that the often-heard argument that the breaker points must be eliminated to have a truly electronic ignition system may be valid from the viewpoint of the purist, but is not of overwhelming importance in practice. This is because, once the breaker points have been relieved of the necessity to carry high current, their life span is appreciably increased. Moreover, the average motorist wants the basic benefits of electronic ignition in his older car without the need for major surgery, such as would be involved in converting the distributor to use a magnetic reluctor or optoelectronics. The system described here is inexpensive, easy to install, and has very good performance characteristics.

In Fig. 5-20 power transistors Q1 and Q2, in conjunction with saturable transformer T1 and associated components, comprise an audio-frequency, square-wave inverter. The secondary voltage from this inverter is rectified by a bridge rectifier and filtered by capacitor C1. Approximately 175 volts dc is made available for operation of the energy charging circuit, the main components of which are capacitor C2, diodes CR1 and CR3, and inductor L1. The basic idea is to quickly accumulate a charge in capacitor C2 and, by means of SCR1, to "dump" this charge into the primary winding of the ignition coil. The timing of these events and the distribution to appropriate spark plugs occur in the same manner as in the conventional ignition system. The advantage of the capacitor-discharge ignition system over the older conventional (Kettering) type system is that a really respectable amount of electrical energy is available to fire worn, poorly adjusted, or fouled spark plugs. Of equal importance, the spark-plug firing volt-

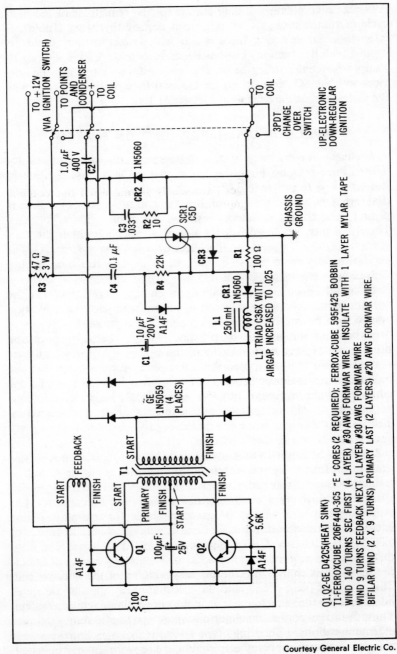

Courtesy General Electric Co.

Fig. 5-20. Capacitor-discharge ignition system.

age not only holds up at high engine speeds with the capacitive-discharge ignition system, it is not readily depleted by low battery voltage when starting a cold engine.

This circuit is similar to most other capacitor-discharge ignition systems in that an SCR is triggered to transfer stored charge from a capacitor to the ignition coil. However, when we investigate the manner in which capacitor C2 is charged, a unique mode of operation becomes apparent. The significant deviations in this circuit are that only 175 volts are developed by the dc-to-dc converter capacitor C1, and the inclusion of inductor L1 and diode CR1. Initial appraisal might erroneously identify this as a low-energy system, since 175 volts and a 1-μF storage capacitor apparently do not compare favorably with other circuits. (This becomes particularly glaring when one contemplates that electrostatic energy storage in a capacitor is proportional to $V^2 \times C$.)

Actually, the accumulation of stored energy is reinforced by so-called dc-resonance charging. This is a technique borrowed from radar modulators in which the voltage impressed across a storage capacitor is *double* the dc voltage of the power supply. When this system is in operation, a series-resonant circuit involving capacitor C2 and inductor L1 tend to produce shock-excited oscillations. Because of "standoff" diode CR1, only the first quarter of oscillation can be sustained. However, this suffices to boost the 175 volts across capacitor C2 by another 175 volts, making the total potential across the terminals of C2 350 volts. Because of the way diode CR1 is polarized, the 350-volt capacitor charge is "trapped" until transferred to the ignition coil when SCR1 is triggered. Triggering of SCR1 is brought about in the following manner. When the ignition breaker points are closed, no path is available from the positive battery terminal to produce the required gate current for triggering. When the breaker points open, the path via resistor R3 is immediately available to charge capacitor C4 and thereby develop a pulse of gate current for the SCR. Subsequent closure of the breaker points then dissipates the charge in C4. The salient feature of this triggering technique is that the ignition breaker points, condenser, and ignition coil can be left as they were in the original system.

When the ignition breaker points open, the SCR triggers and remains conductive long enough to fire a spark plug. When turned on, the SCR is part of a resonant circuit formed by the primary inductance of the ignition coil and capacitor C2. At the outset of the first reverse-voltage alternation of the shocked oscillation, the SCR is turned off. Thus, the circuit is self-commutating. Both commutation and the previously described resonant charging for this capacitive-discharge ignition system work reliably over the entire range of engine speeds in 4-, 6-, and 8-cylinder automobiles.

PULSER FOR SOLID-STATE INJECTION LASERS

Fig. 5-21 shows a system for pulsing the solid-state injection laser with high currents at a low duty cycle. Not only is the circuit strongly reminiscent of gas-laser technology, but it will be seen that high voltage, rather than the several volts ordinarily involved with diodes, is used.

Within dashed area, leads should be kept as short as possible to minimize inductance.

C = 1.0 μF to 100 pF (MYLAR) for desired pulse width.

Courtesy RCA Corp.

Fig. 5-21. Pulser for injection laser devices.

The two series-connected 40852 power transistors provide the path for "quick-charging" capacitor C *between* laser pulses. Although it might initially appear otherwise, the laser diode derives its pulsed energy from the charge stored in capacitor C, *not* directly from the power transistors. Therefore, the capacity of capacitor C determines the *duration* of the current pulse that activates the laser diode.

The 1N914 diodes are connected in the circuit in such a way that the lower 40852 transistor is deprived of forward bias while current is being delivered to the laser diode. But after depletion of the stored

268

charge in capacitor C, the base of this transistor recovers its forward bias and the charging process commences again. Inasmuch as the SCR in series with the laser diode may be pulsed at an audio-frequency rate, capacitor C must be capable of being charged rapidly to provide energy for the laser diode each time that the SCR is turned on.

Since the energy available for the laser diode is proportional to the square of the voltage stored in capacitor C, considerable control of the power residing in the laser beam is obtained from a relatively small variation of the voltage *allowed* to develop across the terminals of capacitor C. This adjustment is provided by the 250K potentiometer.

The capacitor voltage can only attain approximately the voltage at the wiper arm of the potentiometer because the conduction of the 1N3563 diode reduces the forward bias at the base of the output transistor. The charging rate of the capacitor is limited by this action so that the capacitor does charge to a voltage beyond that at the wiper arm of the potentiometer. When the wiper arm of the potentiometer is at its ground end, the capacitor voltage is substantially zero. At maximum adjustment of the potentiometer, the capacitor charges to 500 volts.

The SCR trigger pulses originate in the CA555 IC timer which is connected to operate as an astable oscillator. The repetition rate of the trigger pulses is adjustable by the 1-megohm potentiometer. The output from this square-wave generator is made into narrow, positive-going, unidirectional pulses in the unbiased amplifier stage employing the 2N4036 transistor. Such pulses are more suitable for triggering the SCR than is the original signal.

Typical operating conditions of a family of RCA GaAs injection lasers are given in Table 5-4. The threshold current, I_{th}, is the current at which the device begins *lasing*. At lower currents, the behavior is essentially that of an infrared-emitting diode. Lasing action is characterized by certain desirable optical phenomena, coherent electromagnetic radiation at an extremely narrow bandwidth, and very high power density in the beam.

Other solid-state optical devices can be operated in this circuit, including infrared-emitting diodes and LEDs. (A reduction of the high voltage would generally be necessary.) Also, there are stacked-diode arrangements and diode arrays that can be used with this circuit. Such structures produce more beam power than it is practically feasible to obtain from a single laser diode.

WARNING

Laser Radiation—laser devices in operation produce invisible electromagnetic radiation which may be harmful to the human eye.

Table 5-4. Typical Operating Conditions for Some RCA Injection Lasers

Characteristics at $T_C = 27°$ C and at the Specified Operating Conditions

Total Peak Radiant Flux (Φ_M) (Power Output) Minimum—W	Peak Forward Current (i_F) A	Pulse Duration (t_w) μs	Pulse Repetition Rate (prr) kHz	Typical Threshold Current (i_{th}) A	Typical Peak Forward Drop at i_F (v_F) V	Typical Rise Time ns	Type
1	10	0.2	1	4	4.5	√	SG2001
2	10	0.2	1	3.5	4.5	√	SG2002
3	25	0.2	1	6	6.5	√	SG2003
5	25	0.2	1	6	6.5	√	SG2004
5	20	0.2	1	6	6	√	SG2005
7	40	0.2	1	11	8	√	SG2006
10	40	0.2	1	11	8	√	SG2007
12	75	0.2	1	25	10	√	SG2009
15	75	0.2	1	25	10	√	SG2010
20	100	0.2	1	36	12	√	SG2012

Courtesy RCA Corp.

INDUCTION RANGE

There are two modes of electrical cooking that have enjoyed considerable success in both homes and restaurants. The oldest of these methods is the simple electric range. With this method of cooking, the rather straightforward technique of dissipative heat production in a resistive element is employed. A newer and more sophisticated method is found in the "radar" or microwave range. Here, the heat needed for the cooking process is generated within the food being processed. The temperature rise in the affected food material is the result of dielectric losses, and generally requires a minimal moisture content.

With the advent of suitable solid-state power devices, a *third* method of cooking with electricity has become economically and technically feasible. This is *induction* cooking and involves the production of eddy currents in a metal pot or cooking vessel. The cooking vessel functions essentially as the shorted secondary of a high-frequency transformer. The frequency has to be high relative to the 60-Hz power line frequency in order to produce sufficient heating effect, since the equivalent transformer has an "air core." If the vessel is made of iron or steel, there is additional heating because of magnetic hysteresis. However, best heating stems from the large eddy-current loss in copper vessels.

The "burner" of such an inductive range is the inductive element of a series-resonant *tank circuit*. For the reasons just stated, and for various other practical reasons, the resonant frequency is usually in the vicinity of 25 kHz to 40 kHz. The technique is both electrically and thermally efficient. The "burner," of course, remains relatively cool.

While such a cooking method seemingly has much merit, it has not yet become a commercial success. This may be due to the fact that inductive cooking has appeared on the scene just when the long-heralded impact of microwave cooking was finally materializing. In any event, the circuit of the experimental induction range shown in Fig. 5-22 is bound to stimulate interest.

The heart of the circuit is the SCR series inverter consisting of inductor L1 (the "burner"), resonating capacitor C4, SCR2, diode RD2, and snubbing network R4 and C5. Inductor L1 is spirally wound with several turns of copper tubing to facilitate coupling to the cooking pot. A practical form of this inductor is readily made to have an inductance of about 6 μH. In conjunction with resonating capacitor C4, energy oscillates in these reactive elements and is inductively transferred to the cooking vessel through the physical proximity between the inductor and the vessel. As might be suspected, SCR2 functions as a switch to produce shock excitation of the LC circuit. The oscillatory waveshape is *not* continuous, nor can it be identified as a *damped* waveform. This statement may appear to be contradictory,

271

Courtesy International Rectifier Corp.

Fig. 5-22. Experimental circuit for an induction range.

for most high-frequency electronic apparatus produce *either* continuous or damped oscillations. The operating mode of the inductive range is representative of a third way of "packaging" electrical energy, which is by bursts of integral cycles. Mathematically, it can be shown that such waves are not damped. This is easily confirmed from a practical standpoint, since such waves do not produce harmonics, which are characteristic of all damped wave phenomena. The induction range produces a wave package consisting of *single* complete sine waves. This unusual waveform is shown in Fig. 5-23.

Assume that SCR2 has just been triggered to its on state. The positive excursion of sinusoidal current then takes place in the series-resonant LC circuit. Ordinarily, such a shock-excited LC circuit tends to "ring," that is, generate a decaying wavetrain of oscillatory waves. When the current attempts to reverse in this circuit, the SCR commutates or turns off. However, the reverse current finds a path through diode RD2. Further oscillation is prevented by the off status of the SCR. Thus, only *one* complete sine-wave cycle has been allowed to circulate in the series tank circuit. (To pacify the "purist," it must be conceded that a number of nonideal circuit situations exist and the integrity of the single wave falls short of perfection. For most practical

272

Fig. 5-23. Waveforms produced by
the induction range.

SCR₂ AND DIODE RD₂ CURRENT
50 AMPS/DIVISION
10μs/DIVISION

purposes, however, it is satisfactory to identify it as a single sinusoidal wave of current.)

The gate of SCR2 is repetitively triggered at approximately 22 kHz, or at a frequency about 70 percent of the resonant frequency of the series LC circuit. That this is so can be readily deduced by comparing the "dead time" in Fig. 5-23 with the time required for a complete cycle. This mode of operation is very suitable for the SCR, for it then has more than sufficient time to turn off between cycles.

The triggering circuit is a conventional unijunction transistor relaxation oscillator with an emitter follower as a buffered output. This circuit is powered by a simple half-wave power supply working from a 25-volt isolation transformer.

The dc power supply for the induction range is designed around SCR1, which performs essentially as a controlled output half-wave rectifier. Except for the filter capacitor connected at its output, this circuit is similar to that used in lamp dimmers and in motor speed controls in electric tools. The extra components associated with the gate circuit are intended to reduce the hysteresis which can often be a nuisance with "economy" light dimmers. Such a gate circuit is known as a double-time-constant type. It enables continuous control settings from near zero to maximum output. The rather primitively filtered dc is entirely suitable for shock excitation of the LC resonant circuit. (Of course, the experimenter might find it more convenient to use a large variac and a bridge rectifier for the power supply.)

OPTICALLY TRIGGERED HIGH-VOLTAGE SWITCH

Performing switching operations at high voltages is not always the trivial task we tend to associate with such a "simple" function. Often the requirements are speedy actuation, electrical isolation between the

Courtesy Motorola Semiconductor Products, Inc.

Fig. 5-24. Optically triggered high-voltage switch.

high-voltage circuit and the control circuitry, immunity to false actuation from electrical interference, and "bounceless" closure of the high-voltage circuit. Inasmuch as the switching action is often intended to initiate a precise pulse of energy, electromechanical control techniques are inherently fraught with difficulties. The enthusiastic motivation to do things electronically is often modified by techniques which *combine* electronics with optical technology. The high-voltage switch shown in Fig. 5-24 is a unique example of this evolving design approach.

This high-voltage switch has a myriad of applications such as in lasers, masers, electron microscopes, x-ray equipment, stroboscopy, particle accelerators, copying machines, power supplies, etc. It is a "natural" as a crowbar circuit for protective shorting of a high-voltage source. Although the SCRs tend to latch in their on state, this does not happen in crowbar applications where the voltage of the shorted source is reduced below that needed to supply holding current to the SCRs. Thus, this crowbar action can be self-clearing if the optical triggering signal is momentary, rather than sustained.

The MRD-300 phototransistors and the fiber-optical system constitute the heart of this switching scheme. For all practical matters, it can be stated that these transistors simultaneously receive their optical turn-on signals regardless of the length of the individual light fibers. When these transistors are turned on, they transfer the stored charges from the 0.1-μF capacitors into the gate circuits of the respective SCRs.

The turn-on time of this high-voltage switch is on the order of 300 nanoseconds at a 1-ampere load current. In pulse applications, it can supply 100 amperes to the load for a duration of 4 milliseconds. For best results, the SCRs should be selected to have closely matched rise times.

It is anticipated that this method of triggering series-connected thyristors will prove superior to the older technique of providing each thyristor with its individual pulse transformer.

SIMPLE HALF-WAVE UNIVERSAL
MOTOR CONTROL CIRCUIT

The single SCR phase control circuit shown in Fig. 5-25 is both simple and economical. Nonetheless, such a primitive arrangement will serve many purposes. The neon bulb is one with capability for handling high current pulses.

When capacitor C attains sufficient voltage to ionize the neon bulb, its charge is abruptly depleted as a trigger pulse is delivered to the gate of the SCR. This happens only on the positive excursion of the ac line voltage. During the negative half cycle, the residual voltage in the capacitor is further reduced and both the neon bulb and the SCR gate

Fig. 5-25. Simple SCR universal motor control circuit.

*NE-83, 5AH, A057B, or equiv.
Courtesy RCA Corp.

circuit appear as open circuits. The speed with which the capacitor voltage can build up to the value needed to fire the neon bulb depends upon the setting of potentiometer R2. Therefore, R2 controls the timing or phase of the trigger pulse relative to the power-line voltage wave. This is tantamount to saying that R2 provides adjustment of the conduction time of the SCR and, therefore, of the average current allowed to pass through the universal motor. In this way, the speed of the motor is manually controllable.

This simple circuit does not feature wide-range speed control. The SCR conduction angle may be varied from only about 30 to 150 degrees. Even if full 180-degree conduction of the SCR could be attained, the motor speed would be limited to about 75 percent of its line-voltage speed because of the half-wave operation. Table 5-5 lists the circuit components in Fig. 5-25 for various operating conditions.

Table 5-5. Components for Circuit Shown in Fig. 5-25

AC Supply	AC Current	F_1	CR_1	R_2	SCR_1
120 V	1 A	3 AG, 1.5 A, Quick Act	D1201B	100 K, ½ W	RCA-2N3528
120 V	3 A	3AB, 3 A	D1201B	100 K, ½ W	RCA-2N3228
120 V	7 A	3 AB, 7 A	D1201B	100 K, ½ W	RCA-2N3669
240 V	1 A	3 AG, 1.5 A, Quick Act	D1201D	150 K, ½ W	RCA-2N3529
240 V	3 A	3AB, 3 A	D1201D	150 K, ½ W	RCA-2N3525
240 V	7 A	3 AB, 7 A	D1201D	150 K, ½ W	RCA-2N3670

Courtesy RCA Corp.

HALF-WAVE CONTROLLER WITH SPEED REGULATION

The SCR circuit shown in Fig. 5-26 is also intended for universal motors. Although rivaling the control scheme of Fig. 5-25 in simplicity, this controller provides *speed regulation*. Inasmuch as there are no apparent feedback loops, this may appear surprising. Speed regulation is attained in the following way:

While the SCR is blocking, residual magnetism in the motor causes a counter emf to appear across the motor terminals. This counter emf is proportional to the rotational speed of the armature. In order for the SCR to turn on, the gate voltage at the wiper arm of potentiometer R2 must exceed the counter emf produced by the motor, the voltage drop across diode CR_2 and whatever gate-cathode voltage is needed for triggering. Since this is a half-wave circuit, the SCR can turn on only during the positive half cycles.

Fig. 5-26. Speed-regulated controller for universal motors.

Courtesy RCA Corp.

Suppose that the motor is operating at some nominal speed and its mechanical load is increased. Under such conditions, it is the nature of the universal motor to slow down. However, in this circuit a reduction in the speed of the motor results in decreased counter emf. This, in turn, enables the SCR to fire earlier in the positive half cycle of the power-line voltage wave. As a result, the average motor current increases, thereby causing the motor to increase its speed. The reverse sequence of events applies if the load on the motor is relaxed and the motor attempts to increase its speed. The overall result is that good speed regulation is achieved. The speed regulation is operative not only for variations in shaft loading, but for other disturbances such as erratic brush behavior, varying bearing friction, and temperature effects.

In a circuit of this type, where no RC phase-shifting network is used, the phase delay of the trigger pulse is limited to 90 degrees. In actual applications, the motor cannot be reliably controlled at slow speeds and very light loads. It can be appreciated that there is no counter emf at zero speed. This causes the SCR to fire early in the cycle, thereby accelerating the motor to a speed beyond that corresponding to the setting of R2. The motor must then give up its excess speed via windage and friction. When it has slowed down considerably, the SCR is able to fire again, only to impart another such impulse to the motor. When R2 is set for somewhat higher speed *and* if the motor is a reasonably loaded, such erratic operation, known as "skip-cycling," cannot occur. This is because speed-ups and slowdowns then occur within the "capture range" of the feedback control and the speed is essentially stabilized.

Table 5-6 lists the circuit components used in Fig. 5-26 for various operating conditions.

MOTOR START SWITCH FOR ½-HP
CAPACITOR-START INDUCTION MOTOR

The compelling feature of the induction motor is that its construction is basically simple and it *runs* without the need for brushes or commutators. Unfortunately, the much-used single-phase induction motor is not self-starting. Several starting techniques are employed. One of them connects a capacitor in series with a *starting winding* until the motor attains about 75 percent of full-load speed, at which time the starting circuit is interrupted. This is brought about automatically by means of a shaft-mounted centrifugal switch. Obviously, such an electromechanical device increases the maintenance requirements. The centrifugal switch actually reduces the extremely good long-time reliability that otherwise would be expected from this motor.

Fig. 5-27 shows a scheme for dispensing with the centrifugal switch in induction motors. A ½-horsepower induction motor is automatically taken through its start and run operating modes via a triac. The gate of the triac is connected to the sensing resistance, R1. In this application, resistance R1 is selected to provide gate triggering voltage for motor currents exceeding 12 amperes rms. For this motor, the *peak* starting current is in the vicinity of 40 amperes and the *peak* running current is about 8 amperes. (We are concerned here with *peak,* rather than rms, values because the gate of the triac responds to the peak value of the voltage drop sensed across R1.) It was also

Table 5-6. Components for Circuit Shown in Fig. 5-26

AC Supply	AC Current	F₁	CR₁, CR₂	R₁	R₂	SCR₁
120 V	1 A	3 AG, 1.5 A, Quick Act	D1201B	5.6 K, 2 W	1 K, 2 W	RCA-2N3528
120 V	3 A	3 AB, 3 A	D1201B	5.6 K, 2 W	1 K, 2 W	RCA-2N3228
120 V	7 A	3 AB, 7 A	D1201B	2.7 K, 4 W	500, 2 W	RCA-2N3669
240 V	1 A	3 AG, 1.5 A, Quick Act	D1201D	10 K, 5 W	1 K, 2 W	RCA-2N3529
240 V	3 A	3 AB, 3 A	D1201D	10 K, 5 W	1 K, 2 W	RCA-2N3525
240 V	7 A	3 AB, 7 A	D1201D	5.6 K, 7.5 W	500, 2 W	RCA-2N3670

Courtesy RCA Corp.

Fig. 5-27. Motor start switch for ½-HP capacitor-start induction motor.

Courtesy Motorola Semiconductor Products, Inc.

determined that the maximum starting current occurred about 12 cycles after application of the 60-Hz voltage from the power line. Such start-up characteristics for the motor is with the motor carrying its rated mechanical load.

When the ac line is connected to the motor, the large inrush of current develops a sufficient voltage drop across R1 to trigger the triac into its fully on state of conduction. Under this condition, starting capacitor C2, the starting winding, and the power line are connected in series. Because of the action of capacitor C2, the currents in the starting and running windings are displaced nearly 90 degrees in phase from one another. The motor then behaves as a *two-phase* machine and, accordingly, develops starting torque. (A simple single-phase field simultaneously urges the stationary rotor to turn in both directions and, therefore, produces no starting torque.) As the motor gathers speed, the line current diminishes until the triac no longer receives adequate trigger voltage to keep it turned on. As previously mentioned, the starting circuit is opened at about 75 percent of full-load running speed and the motor thereafter carries its load as a true single-phase machine.

Because of the inductive nature of the starting winding, the snubber network connected across the triac uses a larger capacitor and smaller resistor than is ordinarily employed with noninductive loads.

FULL-WAVE SPEED CONTROL
CIRCUIT FOR ELECTRIC DRILLS

The motor control circuit of Fig. 5-28 not only provides manual speed control, but incorporates feedback that increases available torque when the load becomes greater. In an electric drill, the motor naturally slows down when encountering increased load in an effort to develop higher torque via greater line current. The demand for

NOMINAL R5 VALUES			
MOTOR RATING (AMPERES)	R5		$R5 = \dfrac{2}{I_M}$ $I_M =$ MAX. RATED MOTOR CURRENT (RMS)
	OHMS	WATTS	
2	1	5	
3	0.67	10	
6.5	0.32	15	

Courtesy Motorola Semiconductor Products, Inc.

Fig. 5-28. Full-wave speed-control circuit for electric drills.

more line current is met in this control circuit by automatically increasing the voltage applied to the motor.

Inasmuch as separate access to the armature and field windings is not required, the application is simplified. In many instances, it will be feasible to package the circuit within the tool or appliance itself.

It should be noted that the circuit is similar to that of the basic unijunction trigger circuit shown in Fig. 3-22. The differences involve the addition of circuit components R5, D6, and C2. The objective of the additional components is to make the basic circuit operate so that the conduction angle of the triac is advanced as the motor demands more current. Note that speed regulation is *not* accomplished in this circuit because the sensed quantity, current, is not proportional to speed (as would be counter emf). However, there may be *some* boost in motor speed by virtue of the increased torque produced.

The bridge rectifier consisting of diodes D1, D2, D3, and D4 applies full-wave rectified voltage to the phase-control circuit. Phase control of the triac results from the charging of capacitor C1 through resistors R2 and R3 from the voltage level established by zener diode D5. When capacitor C1 develops sufficient voltage to fire unijunction transistor Q1, the triac receives a trigger pulse through transformer T1. Capacitor C1 quickly discharges through the emitter circuit of Q1.

281

During the time that the triac is *conducting,* the voltage between circuit points A and B is reduced to the extent that the voltage developed across capacitor C1 is not sufficient to fire Q1. The actual voltage *applied* to capacitor C1 then depends upon the motor current because of the sensed voltage drop across sampling resistor R5. Specifically, the greater the motor current, the greater will be the sensed voltage between points A and B and the more of a "head start" capacitor C1 will have when it commences its charging cycle to fire the unijunction transistor. In other words, the triggering time of the triac will be advanced more as the motor encounters greater load and demands higher current. Resistor R5 is chosen so that the voltage applied to capacitor C1 during triac conduction cannot be great enough to break down zener diode D5.

The main difference from a conventional unijunction triggering circuit, the addition of resistor R5, has been accounted for. It is now appropriate to explain the addition of diode D6 and capacitor C2. As previously pointed out, the amount of the preliminary charge imparted to capacitor C1 depends upon the motor current. The net result is that the triggering of the triac is more and more advanced as the motor current becomes greater from heavier mechanical loading. This being so, it can readily be appreciated that Q1 must *not* retrigger while the triac is conducting. However, without diode D6 and capacitor C2, there would be a tendency for such malperformance. The reason for this is that the interbase voltage of the unijunction transistor would then be quite low, making it vulnerable to firing from low emitter voltage. In order to prevent this situation, the interbase voltage is maintained at a high level during triac conduction with the assistance of diode D6 and capacitor C2.

Successful operation has been demonstrated with 2- and 3-ampere motors used in ¼-inch drills. Under these conditions, motor speed has been satisfactorily controlled down to one-third, or less, of maximum speed and the torque enhancement feature has proven its worth. Because of the varying characteristics of different motors, some experimentation may be in order to optimize the value of resistor R5. If this resistance is too high, a surging mode of operation will occur; if it is too low, full advantage will not be obtained from the torque boost.

REVERSING MOTOR-DRIVE CIRCUIT FOR SERVOS AND OTHER APPLICATIONS

A circuit that enables the control of motor speed and direction rotation by means of a dc voltage obviously has many potential uses. Such a circuit is shown in Fig. 5-29. Its operation is based upon the unique characteristics of the diac, triac, bridge rectifier, and a rever-

Courtesy General Electric Co.

Fig. 5-29. Reversing motor-drive circuit for servo systems and other applications.

sible series motor. Although these elements are often encountered in various other circuits, their utilization in this scheme is quite novel.

Consider first the relevant operating features of series motors in which the same current passes through the armature and field windings. This is why the series motor is capable of operation from an ac source, that is, the *relative* torque between armature and field is always exerted in the same direction despite polarity changes. By the same token, the series motor can be reversed by changing the connections of either (but not both) the field winding or armature winding. It is only natural then that a so-called reversible series motor is one with independent access to the armature and field leads.

Now that we have a motor which lends itself readily to both speed control and reversal, it would be desirable to put it through its paces without resorting to mechanical switches. This is where the bridge rectifier circuit enters the picture. When the current source (the triac) provides one polarity, the current path through the bridge is through the pair of diodes appropriately polarized to be forward biased. Of course, current of opposite polarity completes its path through the other pair of diodes. Specifically, positive current out of the triac goes

283

through diode CR6, through the motor armature, through diode CR3, through the field winding and back to the power line. Negative current from the triac goes through diode CR5, through the armature, through diode CR4, through the field winding and back to the power line. Note that switching from one polarity to the other is the same as *interchanging* the armature connections of the motor.

From what has been said, it should be evident that the triac must *not* provide the usual full-wave current which is desirable in many control techniques. In this scheme, it is already obvious that the triac must be able to provide *either* positive or negative current pulses, but *not both*. If "normal" triac operation were used, the motor would be urged to turn one direction during the first half of the ac cycle and the opposite direction during the subsequent half cycle. Over a full ac cycle, the average torque would be zero and much heat and noise would be forthcoming but the motor shaft would not turn. Inasmuch as the triac can be triggered by gate signals of either polarity, it might appear difficult to make it deliver unidirectional current pulses at will. This problem, however, is neatly solved by the diac and its associated circuitry.

The diac is, of course, bilateral in that it will respond to either a positive or negative voltage of sufficient magnitude. If the diac is connected in a phase-shift circuit with a provision for mixing a dc voltage with the ac, the bilateral action of the diac circuit will be altered. Specifically, the diac can be made to respond to *either,* but *not* both, the positive or negative half cycle of the ac wave. This is what has been done in the diac circuit of Fig. 5-29. As a result of the dc control signal, capacitor C2 starts its charge cycles from a definite dc level rather than from near zero as in most triac control circuits. If, for example, capacitor C2 is initially several volts positive because of the dc control signal, the diac will be enabled to fire on positive half cycles *only*. The reverse situation prevails with a negative dc control signal. Diodes CR1 and CR2 are "steering diodes" that provide the needed paths for the dc control signals, but isolate the ac from the dc control terminal. The overall effect of this circuitry is that the triac is operated in the *half-wave,* rather than the more common full-wave, mode. (The control range and maximum speed limitations generally associated with half-wave motor control systems, therefore, prevail with this scheme.)

Variable resistance R1 provides adjustment of the ac voltage available to the phase-control network. By reduction of this voltage, a deadband is imposed in the control characteristic. This is necessary to ensure that full-wave operation of the triac will never occur. On the other hand, the *smaller* the deadband, the more suitable the control is for servo operation. Balance control R4 adjusts the symmetry of the control characteristic with respect to a zero control signal. It compensates for whatever deviations from balanced control characteristics that may accumulate from tolerances in the diac, triac, and motor. Variable resistance R6 is the counterpart of the phase adjustment in ordinary phase-control systems. In this circuit, its effect is suggestive of a "gain" control inasmuch as the basic speed-control function is under command of the dc control signal. All three controls interact somewhat and careful adjustment is initially required to establish desired operating conditions.

STATIC SWITCHES

The static switch is closely related to the solid-state relay, but tends to be a much simpler circuit configuration. This is because the static switch usually does not have the additional components found in solid-state relays. Thus, the static switch generally will not incorporate opto-isolators, zero-crossing circuitry, output rectifiers for dc loads, or various logic provisions. However, static switches with considerable sophistication are coming into prominence and it is no longer entirely clear whether "solid-state relay" or "static switch" is the proper nomenclature in some instances. The static switch is sometimes referred to as a *contactor*.

Fig. 5-30. A simple triac static switch.

	120 VAC, 60 Hz	240 VAC 60 Hz
R	1K, ½ W	2K ½ W
Y	RCA 40429	RCA 40430

Courtesy RCA Corp.

The best way to gain insight into the nature of solid-state switches is to investigate some typical applications. The simple triac static-switch circuit in Fig. 5-30 is widely used. Its main function is to turn on and off a high-current load by means of a small switch with a relatively low current rating.

The circuit shown in Fig. 5-31 is the SCR equivalent of the triac static-switch circuit in Fig. 5-30. Diode D2 provides triggering current for the gate of SCR Q1 and diode D1 serves the same function for the gate of SCR Q2 during the alternate half cycle. Resistor R1 limits the gate current to both SCRs.

Three other simple static-switch circuits are shown in Fig. 5-32. The scheme depicted in Fig. 5-32A is similar to the triac circuit al-

Fig. 5-31. Static-switch circuit using two SCRs.

ready considered in Fig. 5-30, except that a reed relay is used to switch the gate current. This minor expedient provides a very worthwhile feature—the electrical isolation of the gate circuit. Such isolation is desirable in terms of safety and control flexibility. Otherwise, grounding conflicts can produce various malfunctions and failure modes. The three-position control circuit in Fig. 5-32B is interesting in that the diode in position 2 causes the triac to deliver only half-wave pulses to the load. The circuit action is the same as if one of the SCRs in Fig. 5-31 was removed. Thus, the triac is caused to operate as an SCR and the load power in position 2 is one half of that obtained in position 3.

(A) Isolation of the gate circuit by means of a reed switch.

(B) Three-position control provides off, half-power, or full-power modes.

(C) Using a small transformer for gate-circuit isolation.

Courtesy General Electric Co.

Fig. 5-32. Additional triac static-switch circuits.

The circuit shown in Fig. 5-32C is an unusual method for producing on-off load control with gate-circuit isolation. When the switch is open, the secondary winding of the transformer (in this case, the 120-volt winding) has a high impedance and, therefore, does not allow sufficient gate current to fire the triac. However, when the switch is closed, the secondary winding "sees" the reflected short circuit of the primary winding. The impedance of the secondary winding is, therefore, very low and there is sufficient gate current to turn the triac on.

For all of the simple static switches described here, some experimentation may be in order to determine the optimum value of the gate-current limiting resistance. This resistance will be influenced by the characteristics of the particular thyristor used, by temperature, by line-voltage variation, and by load impedance. For the most part, these simple circuits operate best with essentially resistive loads.

A more sophisticated static-switching system is shown in Fig. 5-33. This circuit permits application and interruption of three-phase power to an induction motor or to appropriately connected and rated lamps or heater elements. This arrangement features single logic-signal control of the three-phase currents and very good isolation between the logic control circuitry and the ac power line. The isolation is provided by the photo-coupled isolators. The three CA3059 ICs are employed here as differential op amps and deliver sustained dc "turn-on" signals to the logic triacs when the dc input logic is high. The logic triacs function essentially as driver stages for the power triacs which actually perform the load-current switching functions.

The ICs have internal dc power supplies, but external filtering is required. This is provided by the three 100-μF capacitors connected between terminals 2 and 8 of the ICs. Although these ICs can provide zero-voltage switching for rfi reduction, they are not connected here to take advantage of this feature. When an inductive load such as a motor is switched, the inductance slows down the rate of load current rise, thereby greatly reducing the chief source of rfi in thyristors. On the other hand, the switching of resistive loads, such as lamps or heaters, *will* be accompanied by bursts of rfi. Inasmuch as the switching will generally be performed at widely spaced intervals, it is reasonable to assume that rfi will not be a problem in many applications.

The terminal labeled REF is an artificial "neutral" established by the 8K resistors for the three-phase power line. This provides proper sequencing of the ac operating power for the three ICs. A 220-volt power line is not required since these ICs are specified for essentially constant characteristics over an ac operating voltage range of 120 to 277 volts.

LIMITED-RANGE THREE-PHASE
CONTROL OF LOAD POWER

At first, the circuit in Fig. 5-34 appears to be a scheme for amplitude control of the triac gate in a way somewhat analogous to the method one might use to control transistor load current by varying the base current. It is true that variable resistance R2 controls the current in the gate circuit of the triac. However, since the triac is

Courtesy RCA Corp.

Fig. 5-33. Triac contactor circuit for three-phase loads.

Fig. 5-34. Simple control circuit for applications where limited control range is allowable.

Courtesy Motorola Semiconductor Products, Inc.

either on or off, the load current is not varied by adjustment of the gate current. What actually happens is that resistance R2 varies the time required for the applied gate voltage to attain triggering level. Thus, the circuit in Fig. 5-34 is a primitive phase-control technique. However, unlike the more common triggering arrangements utilizing RC circuits and triggering diodes, this scheme can only provide load current control between 90 and 180 degrees of each line-voltage alternation. Resistance R1 limits the gate current and its value should be experimentally determined for different triacs. This method of load control is commendable in terms of economy and simplicity for those applications where limited control range is not a disadvantage.

USING A TRIAC TO PREVENT CONTACT ARCING IN A RELAY

Electromechanical relays have the desirable feature that the voltage drop across the closed contacts is a small fraction of that attainable with solid-state devices. However, arcing tends to shorten the life of the contact material and objectionable rfi is often produced in the process. Also, contact bounce sometimes causes additional problems, especially when there are logic circuits in the system. By combining a relay with a triac, the desirable features of *both* devices can be realized in many instances. The cost of the combination can actually be *less* than that for one device alone since the current rating of the relay contacts can be drastically reduced. Also, a heat sink that might be required for a triac alone is not likely to be needed for the relay-triac combination.

The basic idea is to have the triac turn on *before* the relay contacts close and to remain on *after* the relay contacts open. While the relay contacts are closed, the triac is shorted out, thereby developing negligible voltage drop and power dissipation. A circuit arrangement for accomplishing this logic is shown in Fig. 5-35. Note that the main current-carrying terminals of the triac are connected directly across the relay contacts. When switch S1 is closed, the triac is immediately triggered through diode D1. The relay does not actuate immediately because of its solenoid inductance and its mechanical inertia. When the relay finally does respond, its contacts then take over the task of conducting the load current. At the same time, capacitor C1 accumulates

289

Courtesy Motorola Semiconductor Products, Inc.
Fig. 5-35. The use of a triac to prevent contact arcing in a relay.

a charge. When switch S1 is opened, the inherent lags in the relay again prevent immediate response. When the relay contacts finally open, the triac *remains* turned on because of the charge stored in capacitor C1. When the charge depletes itself in the gate circuit, the triac finally turns off. In this sequence of events, the triac turned on *before* the closure of the relay contacts and remained on *after* the opening of the relay contacts. Therefore, the maximum voltage developed across the relay-triac combination is on the order of 1.5 volts. Also, the load is virtually spared the effects of contact bounce.

The closing and opening time lags for the relay specified in Fig. 5-35 are each about 15 milliseconds. With other relays, some experimentation may be necessary for the values of R2 and/or C1. The snubber network, R3 and C2, protects the triac from possible ac-line transients. Because there is no welding or pitting of the contacts, a relay can be selected with one-fifth to one-tenth the current capability that would otherwise be mandatory to provide a good safety margin for the effects of contact arcing.

Index